D0506108

LAKE, MIRE AND RIVER ENVIRONMENTS
DURING THE LAST 15 000 YEARS

PROCEEDINGS OF THE INQUA/IGCP 158 MEETING ON THE PALAEOHYDROLOGICAL
CHANGES DURING THE LAST 15 000 YEARS / BERN / JUNE 1985

Lake, Mire and River Environments

During the last 15 000 years

Edited by
GERHARD LANG
Universität Bern
CHRISTIAN SCHLÜCHTER
ETH-Hönggerberg Zürich

A.A.BALKEMA / ROTTERDAM / BROOKFIELD / 1988

The texts of the various papers in this volume were set individually by typists under the supervision of the editors.

CIP-DATA KONINKLIJKE BIBLIOTHEEK, DEN HAAG

Lake

Lake, mire and river environments during the last 15000 years: proceedings of the INQUA/IGCP 158 meeting on the palaeohydrological changes during the last 15000 years, Bern, June 1985 / ed. by Gerhard Lang, Christian Schlüchter. – Rotterdam [etc.]: Balkema. – Ill.
With index
ISBN 90 6191 849 9 bound
SISO 563 UDC 551.31
Subject heading: geology.

Published by
A.A.Balkema, P.O.Box 1675, 3000 BR Rotterdam, Netherlands
A.A.Balkema Publishers, Old Post Road, Brookfield, VT 05036, USA

ISBN 90 6191 849 9
© 1988 A.A.Balkema, Rotterdam
Printed in the Netherlands

Lake, Mire and River Environments, Lang & Schlüchter (eds)
© 1988 Balkema, Rotterdam. ISBN 90 6191 849 9

Contents

2. Fluvial environments

3. Fluvial environments and palaeohydrology: The state-of-the-art
   Two views from Poland

Lake, Mire and River Environments, Lang & Schlüchter (eds)
© 1988 Balkema, Rotterdam. ISBN 90 6191 849 9

Introduction

This volume is the result of a joint meeting of the "INQUA-Eurosiberian Subcommission for the Study of the Holocene" with IGCP 158 on "Palaeohydrology of the temperate zone in the last 15 000 years". The meeting was organized by the Institute of Botany at Berne University and it was held at Schloss Hünigen near Konolfingen in the Bernese Countryside.

Most of the papers published in this volume were presented orally or during a poster session at the meeting. Contributions on sites visited during the post symposium excursion may be found in the following publications:

(1) Excursion Guide. Symposium in Switzerland 24 June - 1 July 1985 of the "INQUA Eurosiberian Subcommission for the Study of the Holocene" and of IGCP Project 158 "Palaeohydrology of the temperate zone in the last 15 000 years" - Systematisch-Geobotanisches Institut, Universität Bern, 1985, 158 p.

(2) Lang, G., Ed. (1985): Swiss Lake and Mire Enviroments during the last 15 000 years. - 428 p., Vaduz (Cramer).

Unfortunately, the publication of the present volume was unduly delayed, mainly owing to the considerable workload of the junior editor. He apologizes deeply for any inconvenience which may have resulted to any of the contributors. Nevertheless, the last manuscript was received on January 28, 1988 ...

The papers are arranged in three sections: (1) Lake and Mire Environments (IGCP Subproject B) with 10 contributions, (2) Fluvial Environments (IGCP Subproject A) with 9 contributions, (3) two state-of-the-art papers on Fluvial Environments, set apart as a special section for lay-out purposes. A number of colleagues and friends have contributed to an improvement of the manuscripts submitted. We are especially grateful to Dr. Mary M. Boucherle, Martinsville (Indiania USA), for providing linguistic help with some of the texts.

The retyping of the texts in section (1) was done by Mrs. Th. Berger at the Institute of Systematic Botany and Geobotany of the University Berne at the same place the final manuscripts were prepared. The retyping of the texts in sections (2) and (3) and the preparation of the final manuscripts was done at the Institute of Foundation Engineering and Soil Mechanics at ETH: Many thanks to Mrs. Th. Frey and to Mrs. I. Schnider for their skillful and patient help and to the Institute for the generous office support.

We are very happy that the publication is finally available and we conclude with the French saying: Tout est bien qui finit bien.

1 Lake and mire environments

Lake, Mire and River Environments, Lang & Schlüchter (eds)
© 1988 Balkema, Rotterdam. ISBN 90 6191 849 9

Lago Cadagno: An environmental history

M.M.Boucherle
Queen's University Kingston, Ontario, Canada

H.Züllig
Rheineck, Switzerland

ABSTRACT: During the last 1000 years, Lago Cadagno has been altered by several land-slides or avalanches resulting in changes in the lake's biota. Evidence of at least four of these events is reflected by the sudden decrease in the concentration of Clado-cera (Crustacea) remains and bacterial pigments from a sediment core. At the time of each of these events sedimentation was increased so much that both kinds of remains were rare due to dilution by allochthonous material. Lago Cadagno has also experienced two periods of eutrophication. Once as a result of the first landslide/avalanche event and the second time due to human habitation in the lake's catchment. The increased produc-tion resulted in a higher concentration of Cladocera remains and a shift in species and an increase in the concentration of bacterial pigments. The lake was dammed for hydro-electric power use and this perturbation also altered the biota.

1 INTRODUCTION

Lago Cadagno is a small (4.92 ha), mero-mictic, alpine (1921 m altitude) lake with a maximum depth of 21 m located in Southern Switzerland (46°28'N, 8°40'E). This lake is interesting because the concentration of hydrogen sulfide in the hypolimnion makes it a suitable habitat for sulfur bacteria. The lake's water chemistry is controlled by its bedrock of gypsum-containing dolo-mite, that supplies the lake with sulfate. Sulfur bacteria produce unique pigments, most notably the carotenoid okenone, that can be extracted from lake sediments and used to indicate past bacterial popula-tions. The presence of these bacteria in a lake also indicates anoxic conditions with available sulfate in the hypolimnion or monimolimnion (BROWN et al. 1984).

In addition to the bacterial pigments, we also studied Cladocera remains extrac-ted from the sediments of Lago Cadagno. Cladocera remains have been used to pro-vide evidence of historical ecological changes in lakes (FREY 1955, GOULDEN 1964, DEEVEY 1969, HOFMANN 1978 and BRUGAM 1978). Cladocera leave chitinous skeletal remains in lake sediments that are resistant to degradation and are usually identifiable to species. Cladoceran species tend to be restricted to certain habitats, for exam-ple either the littoral zone or planktonic zone of lakes, and for this reason can be used to assess the relative size of the littoral and planktonic zones in a lake through its history (ALHONEN 1970). Clado-cera are also sensitive to the trophic con-dition of the lake and may be used to re-construct past lake trophic status (BOU-CHERLE & ZÜLLIG 1983).

A history of environmental change in La-go Cadagno has been elucidated by analyzing bacterial pigments and cladoceran remains. Lago Cadagno has been subjected to a num-ber of perturbations in the last 1000 years. We have found evidence of at least four landslide or avalanche events that deposi-ted allochthonous material in the lake. Two episodes of eutrophication have also occur-red during this period of the lake's histo-ry. Most recently, about 50 years ago, a dam was built on the lake which raised the water level about 2 m, so that lake could contribute to hydroelectric power genera-tion. All of these disturbance have had an effect on the lake's biota that are reflec-ted by changes in species and in concentra-tion of Cladocera remains and in concentra-tion of the bacterial carotenoids.

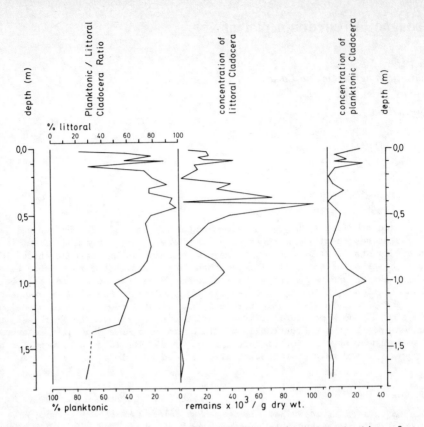

Figure 1. Planktonic/littoral cladoceran ratio, concentration of littoral Cladocera, and concentration of planktonic Cladocera in 180 cm sediment core from Lago Cadagno. Dashed line represents a sample where no remains were found.

2 METHODS

A 180-cm core was raised from a depth of 19.5 cm in Lago Cadagno. This core represents the last 800-1000 years of the lake's history. Cladoceran remains were extracted from dried sediment samples of a known weight by standard methods (BOUCHERLE & ZÜLLIG 1983). Slides were prepared by mounting 0.05 ml of processed sample in Karo mounting medium (TAFT 1978). All identifiable remains were enumerated but only the most abundant skeletal part of each species was included in the final tally. Bacterial pigments and other carotenoids were extracted by methods described by ZÜLLIG (1982) and separated by thin layer chromatography.

3 RESULTS

The data are presented in three figures:

Figure 1 contains the Cladocera planktonic/littoral ratio, the concentration of littoral Cladocera, and the concentration of planktonic Cladocera; Figure 2 illustrates the relative abundance of littoral Cladocera species that made-up at least 5% of the fauna in at least one sample; and Figure 3 depicts the concentration of okenone and other carotenoids in the core. Relative abundance of the planktonic cladoceran species was not portrayed because Daphnia longispina was the only planktonic cladoceran found except in the three uppermost samples where Bosmina longirostris was found. Bosmina made up 71%, 2.9% and 1.8% of the relative abundance of planktonic cladoceran in the 0-2 cm, 2-3,5 cm, 4-6 cm samples, respectively.

Fourteen species of Cladocera were found in the Lago Cadagno sediments including 12 from the Chydoridae, and one each from the Daphniidae and Bosminidae. The members of

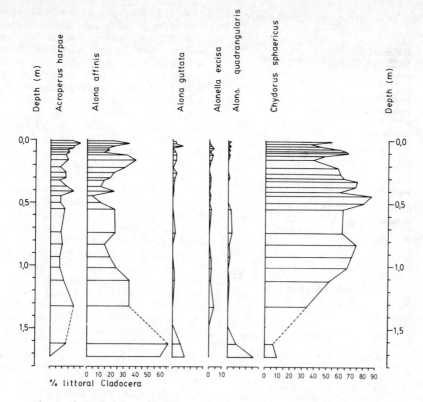

Figure 2. Relative abundance of six chydorid Cladocera from a Lago Cadagno sediment core.

the family Chydoridae are most often littoral-dwellers while the members of the other two families are planktonic. Six of the chydorid species were rare throughout the core and consequently are not illustrated in Fig.2. Of the six chydorid species that were illustrated, three of them made up the majority of the fauna throughout the sediment core.

The oldest sample at 162.5 cm contains a fauna composed of six species, dominated by the planktonic cladoceran Daphnia longispina. No okenone was present among the bacterial pigments at this level. The sample at 147 cm shows that no cladoceran remains were present in the sediment and that pigments were greatly reduced. In the next two samples, 132.5 cm and 111.5 cm, a shift occurs from a cladoceran fauna dominated by planktonic species to one dominated by littoral species. This change was followed by an increase in the relative abundance of Chydorus cf. sphaericus and by a five-fold increase in the numbers of remains at 100

cm. Alona affinis was the dominant chydorid Cladocera prior to the 147 cm sample and afterwards was never more than 40% of the fauna. The bacterial pigments also experienced a huge increase in concentration at this time.

Decreases in the number of Cladocera and in the concentration of bacterial pigments occurred again at 22 cm, 42 cm and 73 cm. In each of these samples both types of remains were greatly reduced compared with samples adjacent to them. This sudden decrease and subsequent recovery is an unusual situation, one that is not normally seen in a stratigraphy.

A second increase in the numbers of Cladocera and the concentration of bacterial pigments took place beginning at 52 cm and continuing until the present. At first glance this does not appear to be true but if the samples where the concentration of remains was reduced to zero are eliminated, the trend of elevated concentrations is easily identified.

5

Concentration of Carotenoids

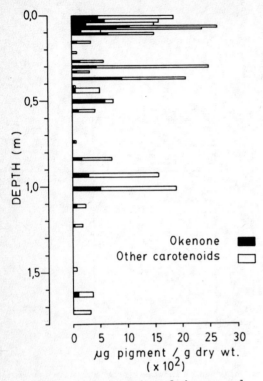

Figure 3. Concentration of okenone and other bacterial carotenoids in a 180 cm sediment core from Lago Cadagno.

4 DISCUSSION

Lago Cadagno has had a less diverse Cladocera fauna over the last 11 000 years than do most temperate lakes. Only 14 species were found in the sediment core. This low number of species may be accounted for by the lake's elevation and consequent climate. This lake is covered by snow for 5-7 months each year. Two other Swiss lakes at nearly the same elevation, Hobschensee and St.Moritzersee, also have a low diversity of species. The paucity of cladoceran species at this elevation is due to the short warm season and perhaps the lack of predators.

In the earliest sediments (800-1000 years B.P.) the cladoceran assemblages and bacterial pigments indicate meromictic and oligotrophic conditions in Lago Cadagno. The planktonic cladocerans outnumbered the littoral cladocerans indicating a large planktonic zone relative to the littoral zone. The first episode of decreased con-

centration of remains (147 cm) was accompanied by a large increase in the amount of quartz in the sample. We believe that an avalanche or landslide was responsible for this situation. Erosion occurring more slowly would not have resulted in the elimination of all organic remains. Further evidence of a large input of allochthonous material is provided by the change in the lake from one dominated by planktonic cladocerans to one dominated by littoral cladocerans (53% at 132.5 cm). This shift in cladoceran fauna signals a change in the relative size of the littoral and planktonic zones. One other explanation for this kind of change in Cladocera would be a decrease in the lake's water level but we have no evidence to suggest that this had happened in Lago Cadagno.

The first episode of decreased concentration of pigments and Cladocera was followed by three similar events. At 22 cm, 42 cm, and 73 cm striking reductions in both types of remains also took place. These reductions are probably the result of a large input of allochthonous material due to a landslide or avalanche. The event at 73 cm appears to have been less severe than the other three because some of each type of remains was found in the sediment, indicating that there was a smaller input of allochthonous material on this occasion.

Lago Cadagno has experienced two periods of elevated production during the last 1000 years. These periods were marked by increases in Chydorus cf. sphaericus, a eutrophic cladoceran, increases in the concentration of all cladocerans, and increases in the concentration of bacterial pigments. The first increase in production (100 cm) followed the first large input of allochthonous material, which decreased the volume of the lake. A shift from a planktonic species dominated lake to a littoral species dominated lake confirms the reduction in lake volume. The lake was probably receiving nutrients at the same rate as before this event, but the smaller volume of the lake led to an increased concentration which resulted in increased production.

The second episode of increased production has occurred more recently beginning at the 52 cm sample. The same kinds of increase in cladoceran remains and pigments occurred as in the first eutrophic episode. But this second episode which continues to the present is probably the result of increased human habitation and agriculture in the lake's catchment. At the top of the core another shift in the planktonic/littoral ratio can be seen. The Cladocera fauna shifts from one dominated by littoral species to one dominated by planktonic species. At the same time there is a change in the planktonic cladocerans from exclusive domination by Daphnia longispina to a dominance by Bosmina longirostris. The concentrations of planktonic cladocerans and bacterial pigments increased rapidly at this time. These alterations in the fauna are coincident with the installation of a dam on the lake. The dam was built to raise the lake and control the water level for hydroelectric power generation.

Lago Cadagno has had an unusual history because large inputs of allochthonous material and the building of a dam have altered the water level. Avalanches or landslides occurring in the lake's catchment have deposited large amounts of allochthonous material that resulted in a change in the relative size of the littoral and planktonic zones, thus, affecting the cladoceran fauna and sulfur bacteria populations and the production of the lake. Each time an avalanche or landslide occurred the allochthonous material deposited diluted autochthonous deposition, so that very few or no organic remains were found in those sediments. Lake production was increased following the first avalanche/landslide event, probably as a result of the lake basin alteration. Human habitation has also effected the lake's water level and production. The water level has been raised 2 m by a dam built about fifty years ago. Production has increased as the number of human inhabitants and agriculture have grown.

5 REFERENCES

ALHONEN,P., 1970: On the significance of the planktonic/littoral ratio in the cladoceran stratigraphy of lake sediments. - Commentationes Biologicae 35, 1-9.

BOUCHERLE,M.M. & H.ZÜLLIG, 1983: Cladoceran remains as evidence of change in trophic state in three Swiss lakes. - Hydrobiologia 103, 141-146.

BROWN,S.R., H.J.McINTOSH & J.P.SMOL, 1984: Recent paleolimnology of a meromictic lake: Fossil pigments of photosynthetic bacteria. - Verh. Internat. Verein. Limnol. 22, 1357-1360.

BRUGAM,R.B., 1978: Human disturbances and the historical development of Linsley Pond. - Ecology 59, 19-36.

DEEVEY.E.S., 1969: Cladoceran populations in Rogers Lake, Connecticut, during late- and post-glacial time. - Mitt. Internat. Verein. Limnol. 17, 56-63.

FREY,D.G., 1955: Längsee: a history of meromixis. - Mem. Ist. Ital. Idrobiol., suppl. 8, 141-164.

GOULDEN,C.E., 1964: Interpretative studies of cladoceran microfossils in lake sediments. - Mitt. Internat. Verein. Limnol. 17, 43-55.

HOFMANN,W., 1978: Bosmina (Eubosmina) populations of Grosser Segeberger See. - Arch. Hydrobiol. 82, 316-346.

TAFT,C.E., 1978: A mounting medium for fresh-water plankton. - Trans. Amer. Micros. Soc. 97(2), 263.

ZÜLLIG,H., 1982: Untersuchungen über die Stratigraphie von Carotinoiden im geschichteten Sediment von 10 Schweizer Seen zur Erkundung früherer Phytoplankton-Entfaltungen. - Schweiz. Z. Hydrol. 44, 1-98.

Lake, Mire and River Environments, Lang & Schlüchter (eds)
© *1988 Balkema, Rotterdam. ISBN 90 6191 849 9*

The Late-glacial and early Holocene history of the vegetation in the Wolbrom area (Silesian-Cracovian Upland, S.Poland)

M.Latałowa
Department of Plant Ecology and Nature Protection, University of Gdańsk, Gdynia, Poland

ABSTRACT: Pollen analysis and ^{14}C datings of the bottom sections of three profiles from the Wolbrom peat-bog are the basis of the reconstruction of the Late-glacial and early Holocene vegetation history. The three cooler and two warmer periods of the Late-glacial are described. The Older Dryas climatic deterioration is discussed on the basis of relative and concentration pollen diagrams. The approximate dates when the several trees spread on to the investigated area in the early Holocene was determined.

1 INTRODUCTION

The peatbog at Wolbrom is a reference site for the Silesian-Cracovian Upland subregion in southern Poland. As there are no lakes or peatbogs in this large area, this particular site is practically the only one where the post-glacial history of the regional vegetation can be studied. For this reason, the Wolbrom peat-bog aroused the interest of palaeobotanists a long time ago. The earliest studies were undertaken by Prof. Jan Trela who, already in 1928, published the first pollen diagrams from this site.

The present paper discusses certain aspects of the Late-glacial and early Holocene history of the vegetation of the Wolbrom area on the basis of materials which will be dealt with in greater detail in a future publication (LATAŁOWA, NALEPKA - in preparation).

2 CHARACTERISTICS OF THE STUDY AREA

The Wolbrom peatbog is located about 200 km south of the most southerly extent of the last glaciation (Fig.1), and lies in an upland area in a depression known as the Wolbrom Gate. This depression is surrounded by flat-topped limestone hills covered with loess derived mainly from the Vistulian glaciation. The highest point in this area is 461 m a.s.l.

Chernozems and brown soils in various stages of dacalcification have formed on the loess. Poorer soils occur chiefly in the river valleys (and also in the Wolbrom vicinity) where the substrate in parts consists of Quaternary sands.

The dominant feature in the present-day landscape of this region, is arable land though originally it was woodland. Only on the limestone rocks and landslip debris did natural sward communities thrive. The commonest woodland communities are at present Fagetum carpaticum and Pino-Quercetum. Tilio-Carpinetum, Phyllitido-Aceretum and Ficario-Ulmetum are rarer (SZAFER 1972). A Vaccinio myrtilli-Pinetum pine forest grows on the sands near Wolbrom.

Lying in the Silesian-Cracovian Upland, the Wolbrom peatbog is adjacent to another subregion, the Miechów Upland. This subregion has been almost entirely deforsted, and the presence of xerothermic sward communities as Inuletum ensifoliae, Seslerio-Scorzoneretum purpureae and Stipetum capillatae is characteristic (SZAFER 1972).

3 DESCRIPTION OF THE SITE

The peatbog, measuring c. 1.5 km by c. 0.5 km, lay at 375 m a.s.l. within a larger area of bogs (about 800 ha), on the watershed between the Biała Przemsza and Szreniawa rivers. The peatbog no longer exists, although it was still possible to make a phytosociological record of it during the 1970s. The state of the vegetation at that time indicated the far-gone degradation of the Oxycocco-Sphagnetea raised-bog communities that had once flourished here (MICHALIK 1976) (Fig.2A).

The **stratigraphy** of this peatbog was

BALTIC SEA

Gdańsk

Szczecin

Noteć

Poznań

Warszawa

Łódź

Pilica

Wrocław

Katowice ●WOLBROM

Kraków

● investigated site

▢ inland dunes

▨ loess

······· furthest extent of the Vistulian ice-sheet /Leszno Phase/

······· Poznan Phase

········ Pomeranian Phase

Figure 1. The location of Wolbrom in relation to the extent of the last glaciation,
loess areas and inland dunes (after
MARUSZCZAK 1983 - slightly simplified).

examined in detail (OBIDOWICZ 1976). The
results are shown in Figs. 2B and 2C which
illustrate the longitudinal and transverse
sections through the deposit and are compared with the description of the sediments from palynologically worked up profiles (Fig.2D).

4 MATERIAL AND METHODS

The present analysis was based on the
bottom part of the previously published
diagram WOL.1 (LATAŁOWA 1976) (Fig.4) and
on two hitherto unpublished diagrams WOL.2
(Fig.5) and WOL.3 (Fig.6). Diagram WOL.2
is the result of a simplified pollen analysis carried out in order to synchronize
the earlier published diagram WOL.1 with
radiocarbon dates. However, because there
are serious distortions in the palynological picture of profile WOL.2 which correspond to the inconsistencies in the ^{14}C
data, this profile is treated merely as
supplementary information.

The WOL.1 profile was obtained in 1969
with the aid of a 5 cm-diameter "Instorf"
corer, WOL.2 in 1972 with a Więckowski
corer and WOL.3 in 1979 with an 8 cm-diameter "Instorf" corer. The localities are
concentrated in the central part of the
peatbog (Fig.2A).

Prior to acetolysis, all samples were
boiled in 10% KOH. Mineral matter was removed with hydrofluoric acid. Only in the
case of profile WOL.3 were 1 cm^3 samples
taken and tablets containing Lycopodium
spores added to them (STOCKMARR 1971) in
order to calculate the sporomorph concentration (Fig.7).

The sum AP + NAP = 100% was taken to be
the basis of all calculations on the percentage diagrams. Sporomorphs of aquatic
and marsh plants were excluded from this
sum.

5 RESULTS

5.1 A simplified description of the sediments

Profile WOL.1

575 - 571 cm   fine sand with admixture
               of silt
571 - 543 cm   sandy silt with traces of
               humus
543 - 530 cm   dark brown, fine sand in-
               terbedded with grey olive
               silt
530 - 481 cm   grey-brown sand with ad-
               mixtures of silt and humus;
               distinctly greater organic
               content at depth 504-500 cm
481 -456,5cm   dark brown, sandy Carex-
               moss peat; strongly decom-
               posed, remains of aquatic
               plants Potamogeton and
               Characeae present
456,5- 375 cm black-brown, Bryalo-parvo-
               caricioni peat, partially
               decomposed with traces of
               sand
375 - 200 cm   black-brown, Carex peat
               strongly decomposed

Profile WOL.2

510 - 496 cm   light grey, unhomogenous
               mud with layers of medium-
               grained sand
496 - 482 cm   grey mud containing sand
               and silt
482 - 472 cm   grey-beige, medium-grained
               sand
472 - 453 cm   grey mud containing sand
               and silt
453 - 372 cm   grey silty sand with layers
               having a higher fine-mine-
               ral fraction content (mud
               and silt)
372 - 348 cm   black-brown, moss - Carex
               peat strongly decomposed
348 - 315 cm   black-brown moss peat,
               strongly decomposed, with

10

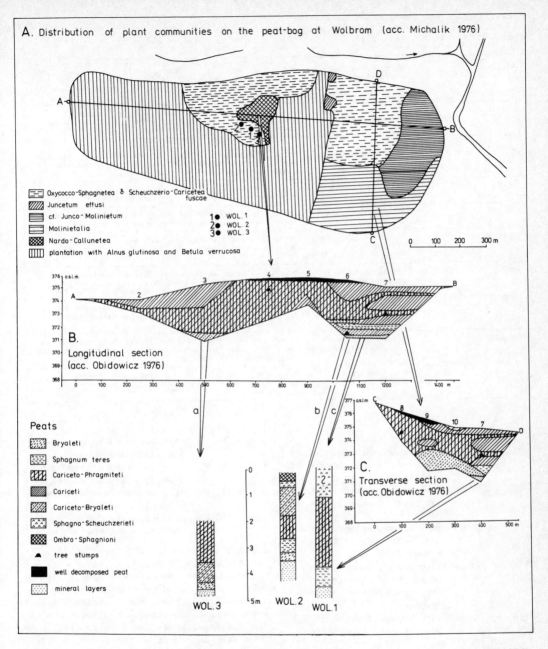

Figure 2. A. Contemporary plant communities on the Wolbrom peatbog (after Michalik 1976) and the position of the localities; B. Longitudinal section through the peatbog (after Obidowicz 1976); C. Transverse section through the peatbog (after Obidowicz 1976); D. Deposits found in the palynologically studied profiles. The arrows a, b, c point to analogies between the deposits in these profiles and deposits examined by Obidowicz (nb – they do not indicate coring locations!).

11

Table 1. WOL.1. Local pollen assemblage zones.

Local PAZ	Depth (cm)	Name of PAZ	Short description of pollen spectra
	573		
WOL_1-1a		Pinus-NAP-Artemisia -Helianthemum	NAP curve (mean 63%) dominates the AP curve. Mean values of Pinus 20,8%, Betula 7%, heliophytes are abundant: Juniperus (mean 3%), Salix (mean 1%), Artemisia (mean 3%), Helianthemum (mean 2%).
WOL_1-1b	----545----	Pinus-NAP-Artemisia- -Cyperaceae	Pinus fall to 5%, Helianthemum, Caryophyllaceae, Chenopodiaceae decrease; at first Gramineae (max. 50%) and later Cyperaceae (max. 75%) rise.
	520 / 511		
WOL_1-2		Pinus-Betula-NAP	Sum of NAP decreases, Pinus curve rises (max. 32%); maximum value of Betula is 14%; heliophytes are abundant: Juniperus (mean 3%), Artemisia (mean 1,5%).
WOL_1-3	----495----	Betula-Juniperus	Juniperus peak appears (max. 5%); NAP curve rises (Cyperaceae max. 40%); Pinus curve declines (↓17%).
WOL_1-4a	----482----	Pinus-Betula-Potamogeton	Sharp rise of Pinus curve (mean 30%); Juniperus decreases (mean 1,5%); culmination of Potamogeton (max. 4%).
WOL_1-4b	----455----	Pinus-Betula-Sphagnum cf. teres (?)	Pinus increasing (mean 53%), decrease in Juniperus, Salix, NAP; Spores of Sphagnum (cf. teres ?) are abundant.
	375		
WOL_1-5		Betula-Larix-Juniperus-Artemisia	Decline in Pinus curve (↓27%), increase in Betula (max. 16%) and NAP (max. 51%; culmination of Larix (max. 4%), Artemisia (max. 4%), Cyperaceae (max. 44%).
WOL_1-6	----355----	Pinus-Filipendula	Pinus is rapidly increasing (↑20%), NAP decreasing, especially the heliophilous taxa. High values of Filipendula (max. 2,5%).
WOL_1-7	----325----	Pinus-Picea-Ulmus- -Corylus-Polypodiaceae	Among tree pollen are dominant: Pinus (max. 78%), Picea (max. 4%), Ulmus (max. 5%), Corylus (mean 3%); Spores of Polypodiaceae are abundant.
WOL_1-8	----230----	Corylus-Quercus-Tilia -Alnus	Rise in the curves of Corylus (max. 12%), Quercus (max. 2,5%) and Alnus; curves of Tilia and Fraxinus are present.
	----198----		

traces of sand

315 - 272 cm dark brown, Sphagnum teres peat, partially decomposed, with traces of sand

272 - 210 cm black-brown Carex-Phragmites peat, partially decomposed

210 - 56 cm dark brown, spongy, Carex-moss peat, partially decomposed

56 - 30 cm dark brown, spongy, Eriophorum-Scheuchzeria palustris peat, partially decomposed

30 - 10 cm dark brown, Eriophorum peat, partially decomposed

Profile WOL.3

495 - 493 cm grey, medium-grained sand

493 - 485 cm grey-brown sandy silt with considerable amounts of plant detritus

485 - 465 cm dark brown, moss peat with sand and silt, strongly decomposed

465 - 435 cm light brown moss peat, slightly decomposed;

435 - 360 cm dark brown Carex-moss peat, partially decomposed; single grains of sand appear from 380 cm downwards

360 - 342 cm dark brown moss-Carex peat, slightly decomposed

342 - 200 cm brown, spongy, Carex peat, partially decomposed

5.2 A brief description of the pollen assemblage zones.

Local pollen assemblage zones (PAZ) have been distinguished in each of the diagrams (Figs.4,5,6,7); their names indicate aspects characteristic of the various sections of the diagrams, while the numbers in the symbols are the result of the search for features common to all three profiles. Tables 1-3 contain the most important information.

The pollen zones are correlated in Fig.3. Analysis of these data shows that equivalent zones are often variously represented in the examined material. This situation is due to the differentation in the local vegetation that grew on the Wolbrom peat-bog and concerns mainly the Late-glacial sections of the diagrams.

Table 2. WOL. 2. Local pollen assemblage zones.

Local PAZ	Depth (cm)	Name of PAZ	Short description of pollen spectra
	484		
WOL-1a		Pinus-NAP--Artemisia	NAP sum (mean 61,6%), Pinus (max. 34%), high values of Artemisia (mean 4,2%).
	------460------		
WOL$_2$-1b		Pinus-NAP-Artemisia--Cyperaceae	Sharp rise of NAP (↑ 35%) - mainly Cyperaceae; Artemisia pollen values are increasing; lateglacial minimum of Pinus (6,2%).
	375		
WOL$_2$-2		Pinus-Betula-Salix--Artemisia	Pinus curve rises up to 30,4%; high values of Salix (max. 6,8%) and Juniperus (max. 4,4%).
	355		
WOL$_2$-3,4,5		Pinus-Betula-Sphagnum cf. teres (?)	NAP curve declines when the Pinus pollen curve rises (mean 68,8%); Sphagnum (cf. teres ?) exceeds 12%.
	265		
WOL$_2$-6,7		Pinus-Picea-Ulmus -Polypodiaceae	Pinus dominates (mean 60%), curves of Picea, Ulmus, Alnus and somewhat later Corylus exceed 1%; spores of Polypodiaceae appear.
	195		
WOL$_2$-8		Corylus-Quercus-Tilia--Alnus	Pinus is decreasing (mean 53%) curves of other trees are rising; Corylus (mean 4%), Quercus (mean 2,5%), Tilia (mean 1,6%), Ulmus (mean 4,3%), Alnus (mean 4,4%).
	85		
WOL$_2$-9		Corylus-Quercus-Ulmus--Sphagnum	Decline of the Pinus curve (mean 18%), culmination of: Corylus (mean 16%, max. 23%), Quercus (mean 9%), Ulmus (mean 8%), Alnus (mean 11,7%)
	17		

Table 3. WOL.3. Local pollen assemblage zones.

Local PAZ	Depth (cm)	Name of PAZ	Short description of pollen spectra
	495		
WOL$_3$-2		Pinus-Betula-B.nana--Potamogeton	Pinus and Betula curves rise; high values of heliophytes: Juniperus (mean 1,6%), Betula type nana (mean 3,4%), Salix t. polaris (mean 1,2%), Artemisia (mean 1,3%); Pediastrum and Potamogeton abundant.
	453		
WOL$_3$-3		Betula-Juniperus-Cyperaceae	Pinus is rapidly decreasing (↓ 35%), whereas NAP (Cyperaceae) curve is rising; Juniperus and Betula type nana values decrease
	447		
WOL$_3$-4		Pinus-Betula--Cyperaceae	The rise of the Pinus pollen curve (mean 54%); heliophytes in small quantities; Cyperaceae are dominant among the herbs (mean 23,6%).
	400		
WOL$_3$-5		Betula-B.nana-Larix--Juniperus-Artemisia	The percentage of Pinus pollen falls (mean 48%); an increase in the percentages of Betula (max. 15%), Larix (mean 1,1%), Juniperus (max. 5%), Betula type nana (max. 2%), Artemisia (mean 1,2%).
	357		
WOL$_4$-6a		Pinus-Filipendula	Pinus is rapidly increasing (↑ 35%), heliophytes are decreasing; Filipendula is abundant (max. 1,2%).
	------343------		
WOL$_4$-6b		Pinus-Betula-B.nana--Polypodiaceae	Pinus and Betula curves cross; Betula type nana is rising (mean 1,2%); Picea and Ulmus exceed 1%; Polypodiaceae spores are very abundant.
	310		
WOL$_4$-7		Pinus-Picea-Ulmus-Co--rylus-Polypodiaceae	Pinus curve dominates (mean 71%); maximum values of other trees are: Picea - 3%, Ulmus - 4%, Corylus 3,2%.
	233		
WOL$_4$-8		Corylus-Quercus-Tilia-Alnus	Curves of Quercus, Tilia, Alnus exceed 1%, Corylus rises (mean 8,8%) and Pinus decreases.
	201		

From the very beginning, the plant communities grew in a mosaic-like pattern; which was a consequence of the uneven bottom of the then-existing water body (Figs. 2B and 2C). Depending on the type of plant community and the level of the water table in particular parts of the peatbog, the deposits built up at different rates; very likely there were also place where the peat-forming processes were periodically retarded. This is reflected in the diagrams by the way in which equivalent pollen zones developed.

Another consequence of the former differentiation in the peatbog vegetation are the differences in the percentage values of the individual components of the pollen zones. Thus, for example, in zone 4 of

13

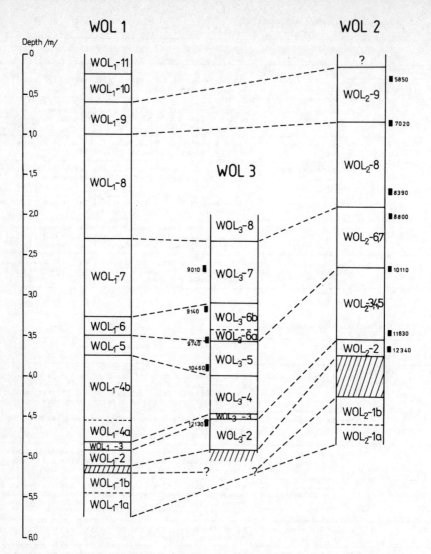

Figure 3. Correlation of local pollen assemblage zones. The synchronic sandy layer is shown hatched.

diagram WOL.1 (Sphagnum teres peat) the NAP percentages are much lower than in the undoubtedly synchronic zone 4 in diagram WOL.3 where an important component of the original communities (moss-Carex peat) were sedges.

5.3 The development of the plant cover
WOL$_{1,2}$ - 1 PAZ
The formation of the Wolbrom peatbog was preceded by a phase when shallow bodies of water were present. Their vegetation was extremely scanty: the pollen diagrams record only single grains of Potamogeton pollen, and there is an almost total lack of plant macrofossils (LATAŁOWA, NALEPKA - in preparation). The very low percentages of tree pollen indicate that the landscape was quite treeless at that time; damp tundra vegetation with bryophytes, Betula nana, Selaginella selaginoides, willows and sedges existed in the immediate neighbourhood of the water bodies. In the dry habitats there were heliophilous communities with juniper, Helianthemum, Artemisia, Ephedra, Sanguisorba minor and others. Pinus cembra was also be found: the presence of Pinus t. haploxylon in the pollen spectra testifies to that.

14

This first stage in the development of the vegetation around Wolbrom, recorded in the diagrams, probably corresponds to the Oldest Dryas. On the diagrams it break down into two distinct parts. The WOL$_{1,2}$-1b PAZ is accompanied in the sediment by fine sand, probably wind-blown, containing a very small number of sporomorphs. This layer serves as a good guide zone for correlating the diagrams.

The Oldes Dryas has also been established in other profiles from Poland (WASYLIKOWA 1964, TOBOLSKI 1966). The differences appearing in the diagrams from these localities are probably the consequence mainly of habitat conditions.

WOL$_{1,2,3}$-2 PAZ
The next phase in the vegetational history of the Wolbrom area, which can be linked with the Bølling, is represented non-uniformly in the diagrams. This is due to the varying rates with which the sediments were deposited. These, in turn, were dependent on the zone of the water body from which cores were taken. Two radiocarbon dates are available for this phase: 12 340 ± 160 (profile WOL.2) and 12 130 ± 160 (profile WOL.3).

During the period under consideration, a water body existed at all three localities. Only in profile WOL.3 towards the end of this phase, are there signs of peatforming moss communities encroaching on this locality. The water body abounded in algae (Pediastrum) and macrophytes like Potamogeton and Sparganium. Patches of damp tundra vegetation with dwarf willows, Betula nana, Arctostaphylos, several species of Gentiana and Saxifraga continued to grow near it.

Birch-pine woodland communities began to expand around Wolbrom at this time, with pine becoming the dominant component, near the end of the period. Aspen and larch also grew in these woods. "Cold steppe" vegetation with Artemisia, Juniperus and Helianthemum still survived in open spaces, though the importance of these communities was clearly diminishing.

The "pine" Bølling has also been recorded at localities in Great Poland (Wielkopolska) (TOBOLSKI 1966, 1985). In the diagram from Witów (WASYLIKOWA 1964), the prevalence of birch is characteristic of the Bølling phase. It is probable that at this time the boundary between pine and birch forest in Poland ran between Wolbrom and Witów.

WOL$_{1,3}$ - 3 PAZ
An almost 30% decrease in the percentage pollen values of pine and a concomitant rise in NAP values (mostly Cyperaceae) are the principal features of this zone, and suggest that the climate was getting worse. This zone could be correlated with the Older Dryas; this suggestion is supported by ^{14}C date 12 130 ± 160 of a layer slightly below the boundary with the previous zone.

This period saw a second expansion of heliophilous communities; evidence for this is the increase in the percentage pollen values and pollen concentrations (Fig.7) of Cyperaceae, Juniperus and a few other species. At the same time, woodlands, whose numbers of pine, larch and aspen were declining, were becoming less important. This is illustrated by both the percentage diagrams and the sporomorph concentration diagrams. The Juniperus peak indicates that the woodland boundary had shifted.

From the pollen diagrams alone, however it is difficult to state definitely to what extent the relatively short-lived climatic oscillation of the Older Dryas affected the transformation of the plant communities. It is likely that the changes recorded in the percentage diagrams and the sporomorph concentration diagram could have been caused by a lowering of the biological vitality of certain species; one manifestation of this may have been the restriction of pine flowering (i.e. PENNINGTON 1977).

During this period the peatbog vegetation had not yet covered the whole area of the water body. Moss communities were already growing at locality 3, but the area around locality 1 was still under water.

As far as diagrams from other parts of Poland are concerned, the Older Dryas oscillation is best reflected in the diagrams from localities in lowland Poland (WASYLIKOWA 1964, TOBOLSKI 1966).

WOL$_{1,3}$ - 4 PAZ
This PAZ illustrates the vegetational changes that took place during the Allerød phase. The zone is in its most complete form in diagram WOL.1 in which two subzones can be distinguished.

During the first phase, represented by the subzone WOL$_1$-4a, birch-pine woodland spread across the area surrounding the Wolbrom peatbog. As the forests expanded, the heliophyte communities gradually dwindled. In the second phase (WOL$_1$-4b)

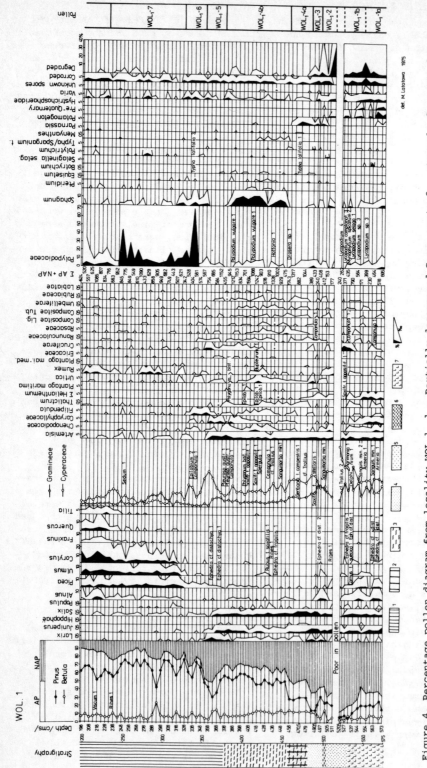

Figure 4. Percentage pollen diagram from locality WOL.1. 1 – Partially decomposed Carex peat, 2 – strongly decomposed Carex peat, 3 – moss and Sphagnum peat, 4 – fine sand, 5 – medium-grained sand, 6 – amorphous humus, 7 – mud and silt.

16

pine woods were predominant. The continual presence of Picea pollen may mean that this tree species grew in the woodlands adjacent to the study site. During this period, communities with Sphagnum teres, typical of the tundra zone, were making inroads on the peatbog (LATAŁOWA 1976).

The expansion of birch-pine and then pine woodland during the Allerød was a ubiquitous phenomenon, recorded as it is on diagrams from southern Poland (SZCZEPANEK 1961, MAMAKOWA 1962), central Poland (WASYLIKOWA 1964, TOBOLSKI 1966), and north-eastern Poland (RALSKA-JASIEWICZOWA 1966, PAWLIKOWSKI et al. 1982). HJELMROOS-ERICSSON (1981) notes that thin birch-pine woodland existed in north-western Poland during the Allerød. Spruce, on the other hand, was at this time present in the Carpathians (KOPEROWA 1962, RALSKA-JASIEWICZOWA 1972) and in upland areas (MAMAKOWA 1962, SZCZEPANEK 1971).

WOL$_{1,3}$-5 PAZ

This zone illustrates the changes that occurred in the plant cover of this area during the next, cool phase of the Lateglacial, the Younger Dryas.

Now the woodland was thinning out again; pine was declining in importance, whereas the opposite was happening with larch. Juniper thickets were becoming widespread. Artemisia pollen attains maximum quantities in this part of the diagrams, so "steppe" communities will have been prevalent.

The vegetation which then was dominant in the Wolbrom area, and in other parts of Poland too, was of the "park tundra" type. At this time only northern Poland was characterised by much larger numbers of juniper and an even sparser covering of pine and tree-birches (HJELMROOS-ERICSSON 1981, LATAŁOWA 1982, PAWLIKOWSKI et al. 1982).

WOL$_{1,2,3}$ - 6,7,8 PAZ

At the start of the Holocene woods with pine, birch and larch again prevailed. Already 10 000 years ago, the importance of spruce was increasing; elm, and later hazel, were making their appearance. Some 8 800 years ago oak, lime and ash began to spread; alder was expanding too.

The post-glacial succession of local peatbog communities started with a short Filipendula ulmaria phase which was followed by the spread of Dryopteris thelypteris, whose spores with their characteristic perine intact frequently turned up in the pollen spectra of all profiles.

Besides Betula nana, Betula humilis also grew on the peatbog (LATAŁOWA, NALEPKA - in preparation), hence the high values of Betula t. nana and Betula pollen in zone 6 of diagram WOL.3.

6 CONCLUSIONS

The diagrams from Wolbrom make an interesting contribution to the discussion on the climate and vegetation of the Late Glacial. The very location of the site, beyond the limits of the last glaciation, in the periglacial zone (Fig.1), provided the theoretical basis for supposing that here a good record of climatic changes, especially during the earlier phases of the Late Glacial, could have been laid down.

On the other hand, however, the specific nature of the Wolbrom peatbog does not render easy the correlation of results obtained from consecutive profiles. As was mentioned earlier, the bottom of the water body exhibited a great diversity of relative depths. This variation had a bearing on which pathways the successions of local communities took at particular spots on the peatbog (Figs. 2A and 2B). The varying rates of accumulation and the diverse composition of pollen spectra of contemporaneous peats such as Bryales peat, Sphagnum teres peat or Carex-moss peat meant that although the borings were made fairly close to one another, the differences in the diagrams are often so great that the profiles seem to have come from entirely different basins. Moreover, the disturbances of the sediment sequence were discovered in profile WOL.2. Apart from the distorted palynological picture, there is here an inversion of ^{14}C dates which is difficult to put down to natural causes; perhaps the peat had been trampled on by animals.

Despite these reservations, the materials presented in this paper do permit a characterisation of the changes in the plant cover of the Wolbrom area and their paleoclimatic interpretation.

The three cool and two warm periods of the Late-Glacial period can be distinguished in these pollen diagrams. Because the boundaries between the periods could not be dated accurately, recourse was had to the terms Oldest Dryas, Bølling, Older Dryas, Allerød, Younger Dryas which refer to biostratigraphic units.

The cool climatic oscillation of the Older Dryas is evident on diagrams WOL.1 and

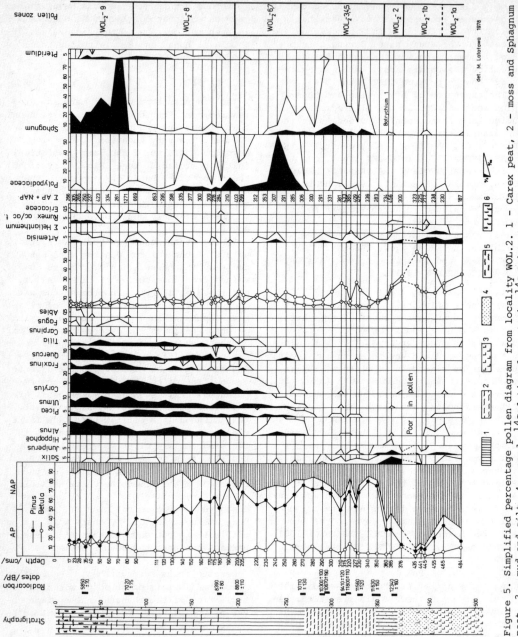

Figure 5. Simplified percentage pollen diagram from locality WOL.2. 1 – Carex peat, 2 – moss and Sphagnum peat, 3 – mud and silt, 4 – sand. 14C dating done at the 14C Laboratory in Louvain (Belgium).

WOL. 3

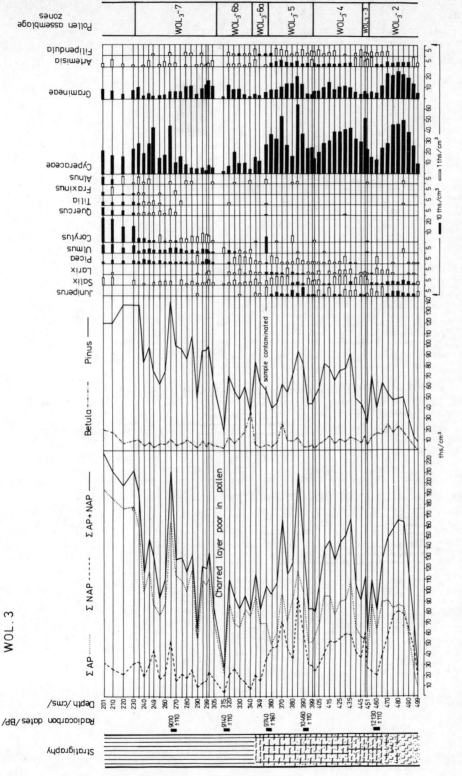

Figure 6. Percentage pollen diagram from locality WOL.3. 1 – Carex peat, 2 – moss and Shagnum peat, 3 – mud and silt, 4 – sand. 14C datings done at the 14C Laboratory in Gliwice (Poland).

19

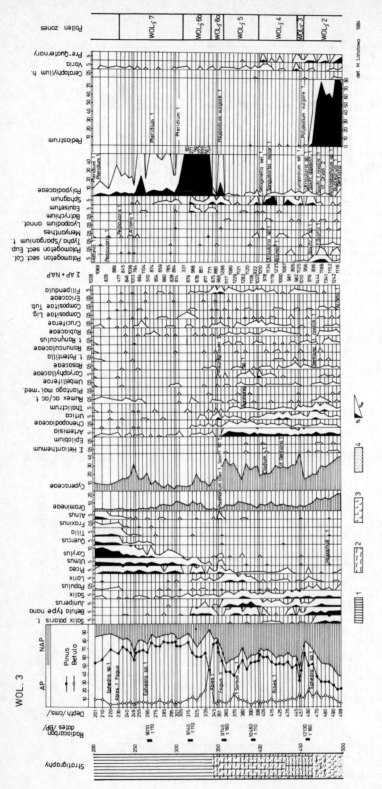

Figure 7. Pollen concentration diagram from locality WOL.3. Description of sediments as for figure 6.

20

WOL.3. The decline of Pinus is contiguous
to the date 12 130 ± 160. The changes in
the percentage pollen diagram are accompanied by a fall in the concentration of
pollen grains, those of pine in particular.

In a discussion on the vegetation and
climate of the Late-glacial WATTS (1980)
stated that very few of the pollen diagrams from Europe reflect unequivocally
the cool climatic oscillation of the Older
Dryas. This cooling is best seen in some
diagrams from Ireland and Great Britain.

It appears that to confirm the Older
Dryas, a site must fulfil a number of fundamental conditions:

1. First of all, the rate of accumulation of sediment should be suitably fast
(over 1 mm per annum) so that a short, approximately 200-year period of time can be
"caught" by the pollen analysis; the standard distance between adjacent samples in
analysed profiles was, until recently, 5 cm.
One has to assume at the same time that
any cooling off in the climate could in
many cases have checked the accumulation
rate of organogenic lacustrine sediments
or of peat deposits.

2. The "response" of the vegetation to a
deterioration in the climate will have been
determined by the stage of development
reached by the plant communities in a given
area. More distinct changes can be expected
near the northern range limits of various
types of woodland communities.

3. Habitat conditions and microclimate
were highly significant as well. It is no
accident that the best Late-glacial pollen
diagrams from Poland have come from areas
of inland dunes (WASYLIKOWA 1964, TOBOLSKI
1966). Any climatic changes in such habitats, "critical" as regards conditions for
survival, must have had a serious effect
on the vegetation there.

The Wolbrom site satisfies the foregoing
conditions. Towards the end of the Bølling
phase, it probably lay close to the northern range limit of pine forests, while the
unstable substrate of sand and loess, affected by the cold and by intensified de-
flation and erosion, may have hindered
plant growth.

7 ACKNOWLEDGMENTS

I wish to express my sincere thanks to
Dr. E.Gilot of the [14]C Laboratory, Univer-
sity of Louvain and to Doc.dr.hab. M.F.
Pazdur of the [14]C Laboratory, Silesian
Polytechnic, Gliwice for providing the
radiocarbon dates.

8 REFERENCES

HJELMROOS-ERICSSON,M., 1981: Holocene development of Lake Wielkie Gacno area,
north-western Poland. - Univ. Lund, Quat.
Geol. Thesis 10, 1-101.

LATAŁOWA,M., 1976: Diagram pyłkowy osadów
późnoglacjalnych i holoceńskich z torfowiska w Wolbromiu (Pollen diagram of the
Late-glacial and Holocene peat deposits
from Wolbrom (S.Poland)). - Acta Palaeobot. 17/1, 55-80.

LATAŁOWA,M., 1982: Postglacial vegetational changes in the Eastern Baltic Coastal Zone of Poland. - Acta Palaeobot.
22/2, 179-249.

MAMAKOWA,K., 1962: Roślinność Kotliny San-
domierskiej w późnym glacjale i holocenie (The vegetation of the Basin of
Sandomierz in the Late-Glacial and Holocene). - Acta Palaeobot. 3/2, 1-57.

MARUSZCZAK,H., 1983: Procesy rzeźbotwórcze
na obszarze Polski w okresi ostatniego
zlodowacenia i w holocenie. - In: KOZ-
ŁOWSKI,J.K. & S.K.KOZŁOWSKI (eds.) Czło-
wiek i środowisko w pradziejach. 32-42,
PWN, Warszawa.

MICHALIK,S., 1976: Roślinność torfowiska.
- In: M.LATAŁOWA, Diagram pyłkowy osadów
późnoglacjalnych i holoceńskich z torfowiska w Wolbromiu (Pollen diagram of the
Late-glacial and Holocene peat deposits
from Wolbrom (S.Poland). - Acta Palaeobot. 17/1, 55-80.

OBIDOWICZ,A., 1976: Geneza i rozwój torfowiska w Wolbromiu (Genesis and development of the peat-bog at Wolbrom (S.Po-
land). - Acta Palaobot. 17/1, 45-54.

PAWLIKOWSKI,M., M.RALSKA-JASIEWICZOWA, W.
SCHÖNBORN, E.STUPNICKA & K.SZEROCZYŃSKA,
1982: Woryty near Gietrzwatd, Olsztyn
Lake District, NE Poland - vegetational
history and lake development during the
last 12 000 years. - Acta Palaeobot.
22/1, 85-116.

PENNINGTON,W., 1977: The Late Devensian
flora and vegetation of Britain. - Phil.
Trans. R. Soc. Lond. B. 280, 247-271.

RALSKA-JASIEWICZOWA,M., 1966: Osady denne
Jeziora Mikołajskiego na Pojezierzu Mazurskim w świetle badań paleobotanicznych (Bottom Sediments of the Mikołajki
Lake (Masurian Lake District) in the
light of palaeobotanical investigations).
- Acta Palaeobot. 7/2, 1-118.

STOCKMARR,J., 1971: Tablets with spores
used in absolute pollen analysis. -
Pollen and Spores 13, 615-621.

SZAFER,W., 1972: Szata roślinna Polski
niżowej. - In: SZAFER,W. & K.ZARZYCKI

(eds.) Szata roślinna Polski. vol. 2, 17-188, PWN, Warszawa.

SZCZEPANEK,K., 1961: Późnoglacjalna i holoceńska historia roślinności Gór Świętokrzyskich (The history of the Late-Glacial and Holocene vegetation of the Holy Cross Mountains). - Acta Palaeobot. 2/2, 1-45.

TOBOLSKI,K., 1966: Późnoglacjalna i holoceńska historia roślinności na obszarze wydmowym w dolinie środkowej Prosny (The Late-glacial and Holocene history of vegetation in the dune area of the middle Prosna valley). - Prace Kom. Biol. PTPN, Wydz. Mat.-Przyr. 32/1, 1-69.

TOBOLSKI,K., 1985: The Bølling flora at Żabinko in the vicinity of Poznań. - Abstracts of papers and posters, Symp. in Switzerland 24 June - 1 July 1985,

p.45.

TRELA,J., 1928: Torfowisko w Wolbromiu (wyniki analizy pyłkowej) (Die pollenanalytische Untersuchung des Torfmoores bei Wolbrom in Mittelpolen). - Acta Soc. Bot. Pol. 5/3, 337-351.

WASYLIKOWA,K., 1964: Roślinność i klimat późnego glacjału w środkowej Polsce na podstawie badań w Witowie koło Łęczycy (Vegetation and climate of the Late-Glacial in Central Poland based on investigations made at Witów near Łęczyca). - Biul. Perygl. 13, 261-417.

WATTS,W.A., 1980: Regional variation in the reponse of vegetation to Lateglacial climatic events in Europe. - In: LÖWE, J.J., J.M.GRAY & J.E.ROBINSON (eds.) Studies in the Lateglacial of North-West Europe, 1-21. Perg.Press., Oxford.

Lake, Mire and River Environments, Lang & Schlüchter (eds)
© 1988 Balkema, Rotterdam. ISBN 90 6191 849 9

Contribution à l'histoire postglaciaire des deux lacs de Clairvaux (Jura, France): Recherches palynologiques et sédimentologiques

M.Magny & H.Richard
Laboratoire de Chrono-Ecology, Besançon, France

ABSTRACT: Several sequences of lacustrine sediments from both "lacs de Clairvaux" (Jura, France), are analysed. The investigation of the history of the regional vegetation is based on pollen analysis; the human impact on the vegetation is considered. The investigation of the water-level changes is based on sediment analysis. Archaeological layers and C 14 datings provided an accurate time scale for the study.

1 INTRODUCTION

Les deux lacs de Clairvaux occupent la petite reculée du Piley qui échancre le plateau calcaire de Champagnole et débouche dans la vaste dépression de la Combe d'Ain (Jura, France). Il s'agit de deux lacs de barrage morainique, qui s'étagent de 527 m (Petit Lac) à 525 m d'altitude (Grand Lac); ils sont par ailleurs caractérisés par une stratification thermique bien marquée en été, par un brassage des eaux en automne et au printemps (Fig.1). La végétation actuelle propre aux deux lacs présente les groupements typiques rencontrés dans la plupart des lacs jurassiens, avec du bord vers le large: phragmitaie, scirpaie, nupharaie, potamaie et enfin zone profonde à Chara et Nitella; chacune de ces parties peut être plus ou moins développée. La chênaie-charmaie, végétation potentielle de cette région, n'existe aux environs des lacs que très morcelée: elle a été transformée par des enrésinements importants ou remplacée après défrichements par des cultures et des prairies.

Les rives du Grand Lac, plus particulièrement à l'extrémité Nord, ont été occupées entre 3600 et 1700 B.C. par des habitats lacustres (Néolithique et Bronze ancien). Connus depuis 1870 ces habitats préhistoriques sont l'objet de nouvelles fouilles archéologiques depuis 1970 sous la direction de P.Pétrequin. Les recherches palynologiques et sédimentologiques exposées ici avaient pour but de préciser l'environnement naturel dans lequel s'étaient développés ces habitats et de montrer quels

avaient pu être les rapports entre l'homme et le milieu.

2 RECHERCHES PALYNOLOGIQUES

Les principales zones d'habitats fouillées au bord du Grand Lac on fait l'objet d'analyses palynologiques (H.RICHARD, 1983). Comme toujours lors des analyses effectuées sur l'emplacement d'un habitat, les résultats sont marqués par une très forte proportion de pollens d'herbacées. Il est maintenant reconnu (H.RICHARD, 1984) qu'une partie importante de ces pollens d'herbacées - en particulier les pollens de céréales - est due à un apport anthropique lié, entre autres causes, au stockage de fourrage et au battage ou au vannage des céréales sur le site. Afin d'éviter ce type de pollution, une analyse pollinique a été faite au bord du Petit Lac, soit à environ 1 km de l'habitat le plus proche (Fig.2).

Ce qui frappe d'abord, c'est la proportion très élevée de pollens d'arbres (AP) dont le pourcentage ne descend jamais endessous de 85% alors qu'il était parfois inférieur à 20% dans certains niveaux attribués au Néolithique final du Grand Lac. Au Petit Lac, de tels pourcentages ont plusieurs causes:
- la proximité des zones à vocation forestière; les pentes très fortes entourant le lac n'ont jamais pu être transformées en cultures ou en pâtures;
- un apport, par les ruisseaux d'alimentation, de pollens d'arbres provenant des massifs forestiers du plateau supérieur;

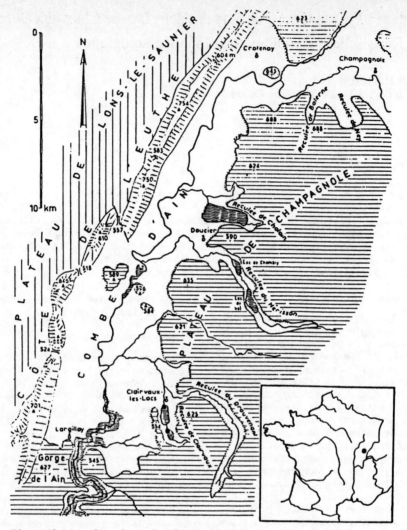

Figure 1. Les deux lacs de Clairvaux dans la Combe d'Ain (d'après M. Campy, 1982). Les deux points noirs indiquent l'emplacement des sondages.

- les zones défrichées étaient limitées et surtout parfaitement choisies en fonction du sol et de la proximité des habitats.

Le diagramme est marqué par une lacune visible dans la variation brutale de certains pollens (Fagus, Ulmus, Tilia, Abies, Alnus entre autres), elle peut se situer entre 3300 et 3000 B.C. L'analyse précise des microvariations du rapport AP/T montre deux phases:

- avant le début du Subboréal, les augmentations des pollens d'herbacées sont ponctuelles, limitées dans le temps:
  - depuis le passage Atlantique ancien-Atlantique récent - soit aux environs de 4000 B.C. - jusqu'à la lacune - soit aux environs de 3300 B.C. - elles apparaissent très locales, liées aux zones humides entourant le Petit Lac (par exemple la hausse des cypéracées à 370 cm de profondeur).
- depuis la lacune jusqu'au passage Atlantique-Subboréal, elles sont beaucoup plus liées à des défrichements qui entrainent une hausse des graminées, d' Artemisia, de Plantago lanceolata et la présence plus constante des Chénopodiacées, des Urticacées.
- à partir du début du Subboréal, le pourcentage des herbacées augmente très pro-

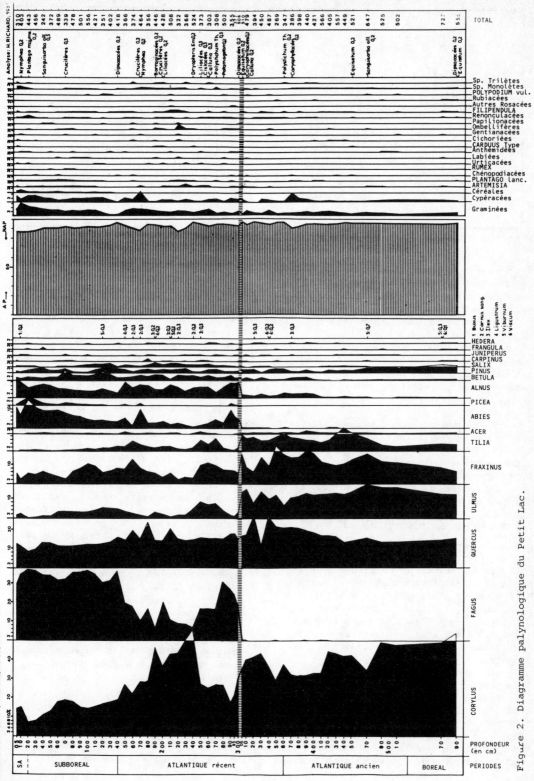

Figure 2. Diagramme palynologique du Petit Lac.

25

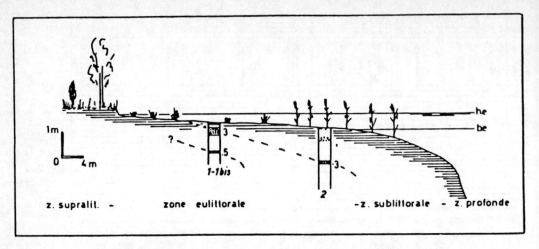

Figure 3. Profil stratigraphique des sondages du Petit Lac.

gressivement, aucun retour des pollens d'arbres n'interrompt cette augmentation, et surtout, les courbes des céréales, d'Artemisia, de Plantago lanceolata sont continues,traduisant la présence constante de l'homme.

Cette analyse du Petit Lac de Clairvaux montre donc deux types d'occupations: d'abord épisodiques pendant l'Atlantique récent puis, à partir du début du Subboréal, une sédentarisation marquée par l'exploitation plus importante du territoire. Ce type d'analyse peut être interprétée en fonction des variations des habitudes agricoles: longues jachères en forêt au Néolithique moyen faisant place à des jachères beaucoup plus courtes liées à une organisation différente du terroir au Néolithique final-Bronze ancien.

3 RECHERCHES SEDIMENTOLOGIQUES

3.1 La méthode

En relation avec les problèmes afférents à l'interprétation des habitats préhistoriques ("Pfahlbauprobleme") et dans une perspective plus générale d'étude de l'évolution postglaciaire de deux bassins lacustres, les analyses sédimentologiques ont été axées avant tout sur la mise en évidence d'éventuelles fluctuations lacustres.

Une première étape a donc consisté à rechercher des critères pour apprécier la profondeur de dépôt des sédiments des carottes analysées, en constituant des "modèles". Le terme de modèle s'entend ici comme l'élaboration d'un ensemble de paramètres qui servent de références dans l'étude et l'interprétation de séquences holocènes.

La constitution de ces modèles s'appuie sur l'étude des sédiments que l'on rencontre actuellement dans les lacs jurassiens, selon des transects perpendiculaires à la rive et englobant littoral, beine et talus. Ainsi ont été reconnus des critères géochimiques et granulométriques, ainsi que d'autres concernant les différents éléments composant un sédiment (décompte à la loupe binoculaire). L'un des critères les plus pertinents est sans doute la représentation de chacun des types de concrétions carbonatées qui obéissent à une zonation parallèle au rivage (M.MAGNY, 1984). La chronologie a été définie en référence à la palynologie et à l'archéologie. Les différentes données sédimentologiques sont réunies sous la forme de diagrammes.

Sur la base de cette approche sédimentologique a été menée une étude simultanée des remplissages du Petit et du Grand Lacs:
- dans l'état actuel de notre documentation aucun habitat préhistorique ne s'est installé sur les bords du Petit Lac où les séquences holocènes sont donc restées à l'abri de toute pollution anthropique jusqu'à une époque très récente.
- l'étude des stratigraphies mises au jour sur le site archéologique de la Motte au Magnins (Grand Lac) a privilégié les couches de sédiments lacustres déposés avant ou entre les différentes phases d'habitat.

3.2 Les résultats

En ce qui concerne le Petit Lac, plusieurs épisodes ont pu être mis en évidence (Fig. 3, 4 et 5):
- Atlantique ancien: le niveau du lac est voisin de l'actuel (épisode 1),

- début de l'Atlantique récent: le lac connaît une régression (épisode 2),
- au cours de l'Atlantique récent: le niveau remonte au-dessus de la cote moyenne actuelle (épisode 3),
- dans une première partie du Subboréal: le lac subit une nouvelle régression plus forte que la précédente (épisode 4),
- dans une deuxième partie du Subboréal, ou déjà au cours du Subatlantique: le niveau remonte au-dessus du niveau moyen actuel (épisode 5),
- au cours du Subatlantique: le lac retrouve son niveau moyen actuel (épisode 6).

S'agissant du Grand Lac de Clairvaux, la séquence étudiée sur le site de la Motte aux Magnins fait état elle aussi de plusieurs fluctuations qui pourraient être reliées comme suit aux précédentes:
- les sédiments antérieurs à une couche archéologique datée de 2900 B.C., montrent la succession d'un haut niveau, d'une régression et d'un haut niveau que l'on pourrait corréler avec les épisodes 1, 2 et 3 du Petit Lac.

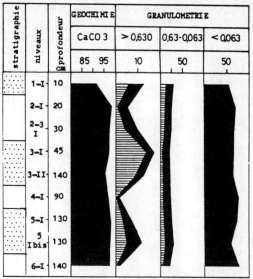

Figure 4. Diagramme sédimentologique du Petit Lac. Géochimie et granulométrie. Pointillés: craie granuleuse. Hachures: fraction supérieure à 2 mm ou fraction supérieure à 0,2 mm.

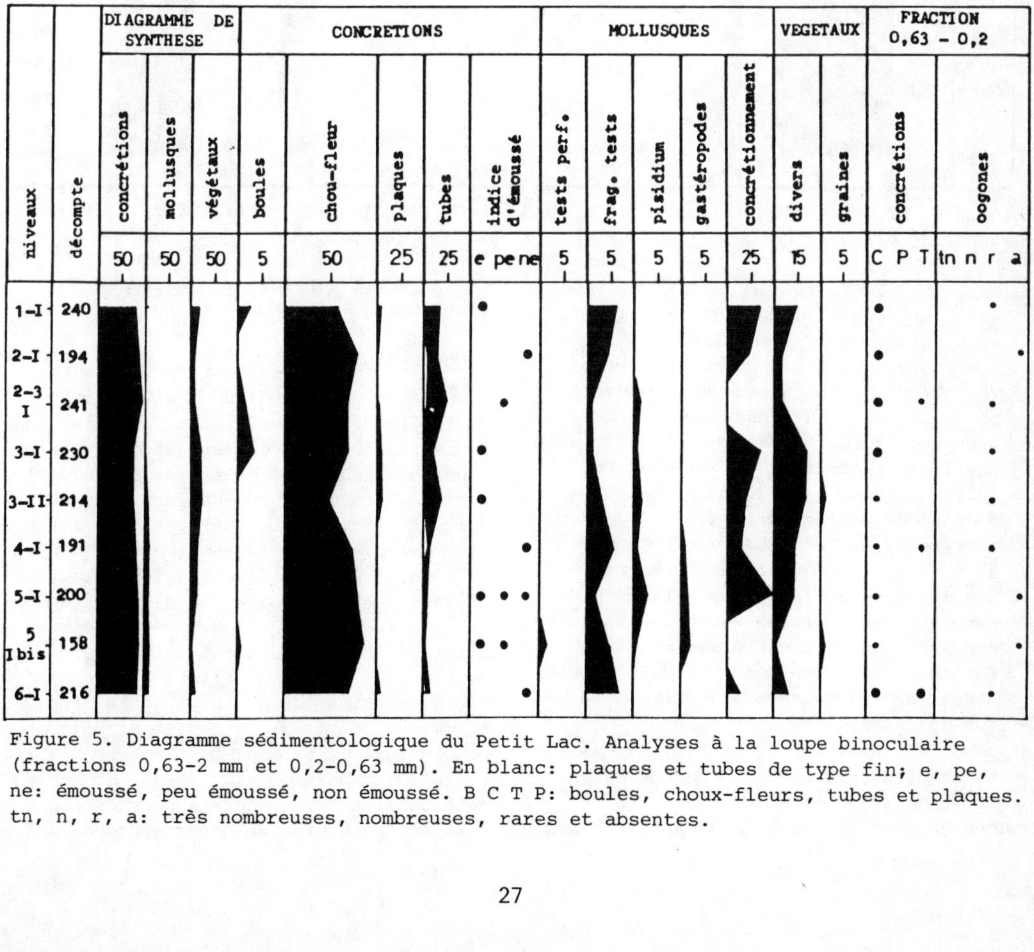

Figure 5. Diagramme sédimentologique du Petit Lac. Analyses à la loupe binoculaire (fractions 0,63-2 mm et 0,2-0,63 mm). En blanc: plaques et tubes de type fin; e, pe, ne: émoussé, peu émoussé, non émoussé. B C T P: boules, choux-fleurs, tubes et plaques. tn, n, r, a: très nombreuses, nombreuses, rares et absentes.

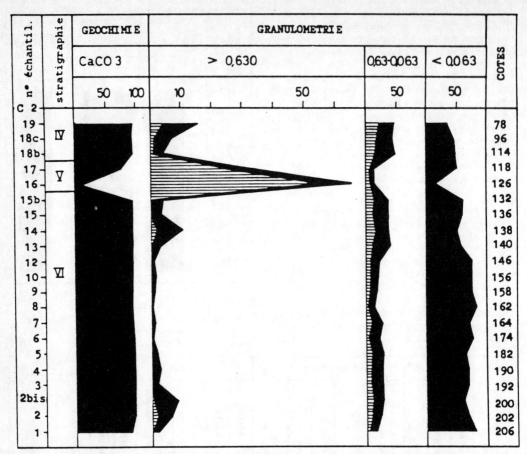

Figure 6. Diagramme sédimentologique du Grand Lac. Géochimie et granulométrie. IV et VI: couches de craie lacustre, V: couche archéologique (2900 B.C.)

- vers 2900 B.C., le lac connaît une nouvelle régression suivie d'une transgression relativement importante, puis d'une régression qui s'amorce progressivement vers 2400 B.C. Le Bronze ancien correspondrait ici à une phase d'atterrissement prononcé (formation d'anmoor), précédant une légère remontée du niveau du lac (niveau actuel). Ce deuxième ensemble d'épisodes postérieurs à 2900 B.C. serait représenté au Petit Lac par les épisodes 4, 5 et 6; la séquence de la Motte aux Magnins apparaît plus complexe et plus complète que celle du Petit Lac en raison vraisemblablement d'une meilleure fossilisation des dépôts lacustres protégés par les couches archéologiques sus-jacentes.

Quant aux habitats lacustres, ils se sont établis soit lors de régressions, soit au début de transgressions. Enfin, aucun habitat ne s'est développé au cours de la régression survenue au début de l'Atlantique récent.

4 CONCLUSIONS

Au terme de cette étude, il apparaît difficile dans l'immédiat de préciser la signification paléoclimatique de ces fluctuations lacustres. On se bornera à souligner l'existence de régressions lacustres dès l'Atlantique. Le Subboréal est caractérisé par plusieurs régressions comme on a pu l'observer pour d'autres lacs. Toutefois un rapprochement entre ces différents épisodes exige l'établissement d'une périodisation très précise qui fait encore souvent défaut. De même, l'interprétation des microvariations de certains pollens en termes de pratiques agricoles si elle semble être confirmée par l'analyse d'autres sites (G.LAMBERT et al., 1983), demande elle aussi à être précisée pour être généralisée. La poursuite des recherches paly-

28

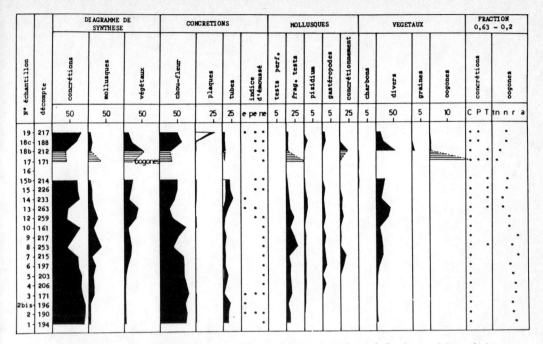

Figure 7. Diagramme sédimentologique du Grand Lac. Analyse à la loupe binoculaire. Excepté les oogones de characées, les végétaux ne sont pas décomptés dans la couche archéologique (apports anthropiques).

nologiques et sédimentologiques, concertées et systématiques, sur les bassins lacustres du Jura et des Alpes du Nord en France, apportera peu à peu des élèments de réponse tout en permettant de mieux cerner l'histoire de l'habitat palafittique en rapport avec l'histoire des lacs. C'est l'objet et le sens du programme du Laboratoire de Chrono-Ecologie de l'Université de Besançon.

5 BIBLIOGRAPHIE

LAMBERT,G., P.PETREQUIN & H.RICHARD, 1983: Périodicité de l'habitat lacustre et rythmes agricoles. - L'Anthropologie. T 87, 3, 393-411.

MAGNY,M. (à paraître): Esquisse d'un modèle pour une approche sédimentologique des habitats lacustres de Clairvaux. In: Les sites littoraux néolithiques de Clairvaux-les-Lacs (Jura). - Le Néolithique moyen 2.

MAGNY,M. (à paraître): Les niveaux VI, V, IV et III de la Motte aux Magnins: l'histoire du Grand lac de Clairvaux du Néolithique moyen au Néolithique final.In: Les sites littoraux néolithiques de Clairvaux-les-Lacs (Jura). - Le Néolithique moyen 2.

RICHARD,H., 1983: Nouvelles contributions à l'histoire de la végétation franc-comtoise tardiglaciaire et holocène à partir des données de la palynologie. - Thèse, Université de Besançon. 155 pp.

RICHARD,H., 1985: Un exemple de pollution anthropique dans les analyses polliniques: les habitats néolithiques du Grand Lac de Clairvaux (Jura). - Actes du Colloque "Palynologie et Archéologie", Valbonne, 279-297.

Lake, Mire and River Environments, Lang & Schlüchter (eds)
© *1988 Balkema, Rotterdam. ISBN 90 6191 849 9*

Diatoms in bottom sediments of Lake Hobschen, Simplon, Switzerland
Preliminary report

B.Marciniak
Institute of Geological Sciences, Polish Academy of Sciences, Warsaw, Poland

ABSTRACT: A preliminary diatom analysis of the lower part of sediments (depth 977.5-895 cm) of the core HO2 from the Lake Hobschen allowed to distinguish six diatom phases (H1-H6) concordant with a lithologic subdivision of sediments. In the lowermost part of the section a very low frequency of diatoms is observed (phase H1-H3). The overlying sediments (depth 942.5-895 cm) contain numerous and highly varying diatom species (phase H4-H6). During the diatom phase H4 a rapid development of the Fragilaria flora occurred. A considerable drop in diatom frequency is noted in the phase H5. A role of the genus Fragilaria (excl. F.alpestris, F.pinnata) decreases whereas the genera Amphora, Pinnularia, Caloneis, Hantzschia, Ceratoneis and Diatoma are more important. In the phase H6 a renewed development of some Fragilaria species occurred accompanied by Navicula, Achnanthes, Synedra and Eunotia.

1 INTRODUCTION

Lake Hobschen is located at Simplon-Pass, 2017 m a.s.l., in the western Central Alps (Wallis). The lake lies above the present tree line and near the upper limit of the alpine tree line. The bedrock is gneiss.

Lake Hobschen HO 2 core is 977.5 cm long and was collected in the deepest part of the lake, at water depth equal 325 cm (Fig. 1). A diatom analysis of this core was possible due to its kind transfer to the authoress by Professor G.Lang from the University of Bern[1].

Up to now a preliminary analysis of 34 samples from the lower part of the sequence (depth 977.5-895 cm) was done. These sediments have a considerably varying lithology: sand, clay, silt, clayey gyttja, clay, gyttja. Within the lowermost sediments at 977.5-945cm, three diatom phases were distinguished (H1, H2, H3). Due to a lack of diatoms in these sediments, lists of taxons were prepared, only some of

[1] Pollen zones and macrofossils of this core were presented by G.LANG and K.TO-BOLSKI (1985) - Late Glacial and Holocene environment of a lake near timberline; Diss. Bot. 87, 209-228.

which are presented in the diagram (Fig.2). On the other hand the three following diatom phases (H4, H5, H6) found in the overlying sediments (depth 942.5-895 cm), possess a richer and more diversified diatom flora. Results of a quantitative analysis of diatoms predominant in this part of the section is presented in Fig.2 and a description of distinguished diatom phases.

2 DIATOM SUCCESSION

2.1 Diatom phase H1 occurs in the lowermost part of the sediments (977.5-972.5 cm). This flora contains single freshwater, mainly cosmopolitan species, oligohalobous (indifferent) and pH circumneutral ones of the genera Navicula, Cymbella, Stauroneis, Nitzschia, Melosira. If these diatoms do not come from a secondary deposit, then they mark the initial developmental stage of the lake. Amidst the fossil Quaternary diatoms, the freshwater allochthonous species cannot be separated from the autochthonous ones. Only an evaluation of diatom content in sediments of the lake substrate and surroundings can be done, particularly if these sediments are of glacial origin. Minerogenic deposits of the Pleistocene age with microfossils (palynomorphs, diatoms, sponge spicules) in a secondary de-

PLATE EXPLANATIONS

Plate I. Figs 1-26 Fragilaria elliptica; Figs 27-31 girdle view of valves; Figs 32-55 Fragilaria elliptica f.minor, 56-60 girdle view of valves (LM, x 1700).

Plate II. Figs 1-9 Fragilaria pseudoconstruens; Fig. 10 girdle view of valves; Figs 11-13 Fragilaria pseudoconstruens var.ibigibba; Fig. 14 girdle view of valves; Figs 15-19 Fragilaria pseudoconstruens var. bigibba f.?; Figs 20-34 Fragilaria pseudoconstruens f.?, (LM, x 1700).

Plate III. Figs 1-30 Fragilaria pinnata; Figs 31-38 Fragilaria pinnata var. lancettula; Figs 39-44 Fragilaria pinnata, girdle view of valves; Figs 45-48 Fragilaria sp.; Figs 53-73 Fragilaria microstriata; Fig. 65 girdle view of valves; Figs 74-75 Fragilaria brevistriata?, (LM, x 1000).

Plate IV. Fig. 1a) Fragilaria elliptica, b) Fragilaria elliptica f. minor; Figs 2-4 Fragilaria elliptica f. minor; Fig. 5 Fragilaria elliptica, exterior view of valve; Fig. 6 Fragilaria alpestris?, exterior view of valve, (SEM, scales 2 um).

Plate V. Figs 1-2 Fragilaria pseudoconstruens, exterior view of valves; Figs 3-4 Fragilaria pseudoconstruens, interior view of valves; Fig. 5a) Fragilaria pseudo-construens, b) Achnanthes borealis; Fig. 6a) Fragilaria pseudoconstruens f.?, b) Fragilaria pinnata, (SEM, scales 2 um).

Plate VI. Figs 1-3 Fragilaria microstriata; Figs 2-3 frustule in girdle view; Figs 4-6 Fragilaria microstriata, exterior view of valves, (SEM, scales 2 um).

Plate VII. Figs 1-2 Fragilaria pinnata, interior view of valves; Figs 3-4 Fragilaria alpestris, exterior view of valves; Figs 5-6 Synedra sp. (SEM, scales 2 um).

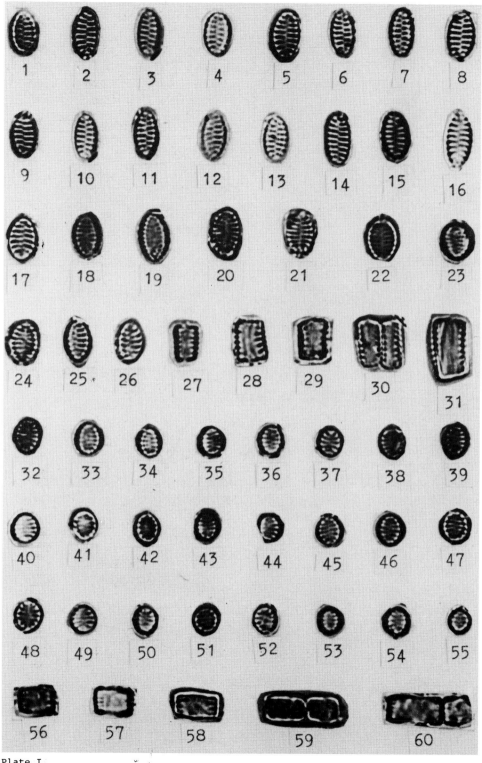

Plate I

Plate II

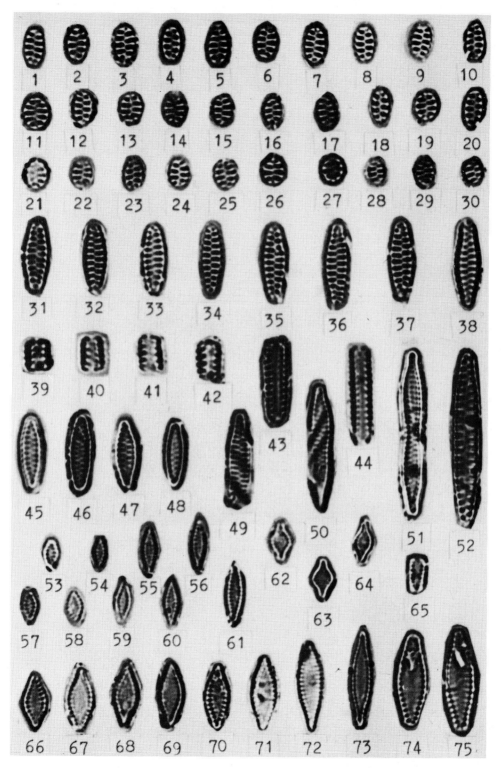

Plate III

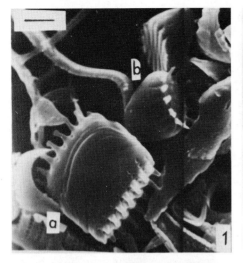

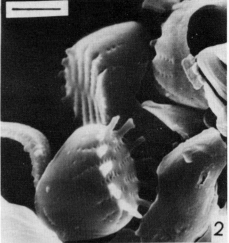

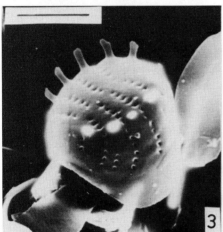

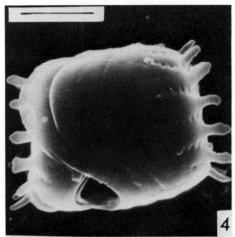

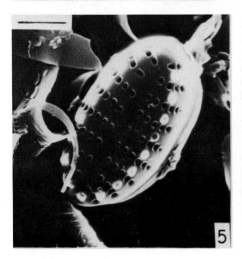

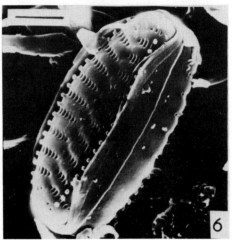

Plate IV

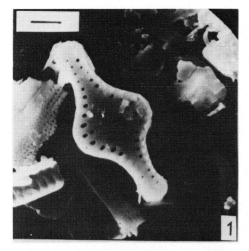

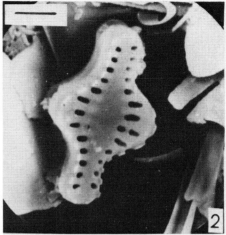

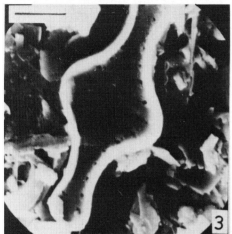

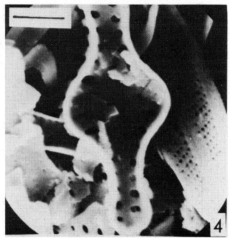

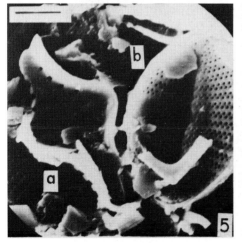

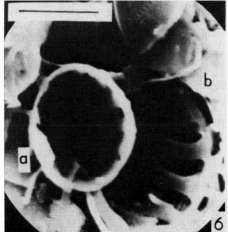

Plate V

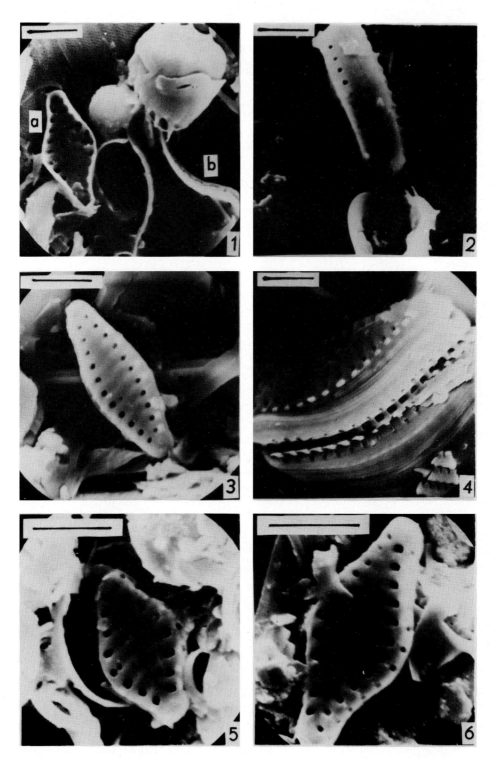

Plate VI

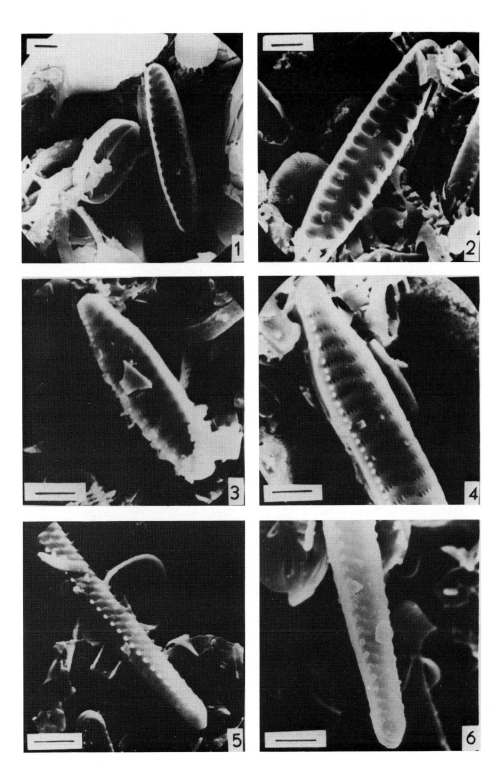

Plate VII

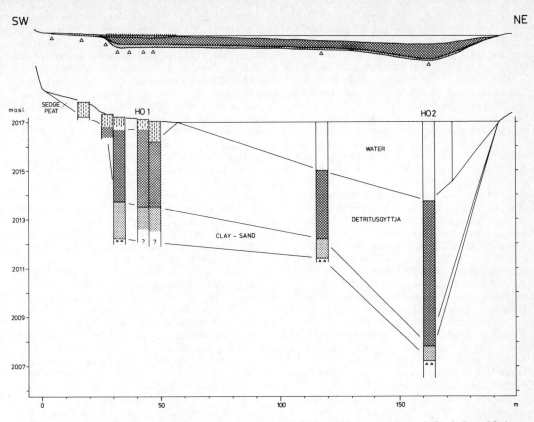

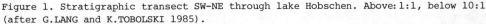

Figure 1. Stratigraphic transect SW-NE through lake Hobschen. Above: 1:1, below 10:1 (after G.LANG and K.TOBOLSKI 1985).

posit, can be used for stratigraphic purposes and to describe these sediments and reconstruct the glacier advance directions (U.MILLER 1982).

2.2 Diatom phase H2, mainly the subphase H2a (depth 970 cm), has more diatoms than the previous phase. Asterionella, Cyclotella, Denticula and Pinnularia are more abundant whereas Fragilaria, Cymbella, Ceratoneis, Navicula, Achnanthes, Amphora, Nitzschia, Stauroneis, Eunotia are more rare. Besides the planktonic diatoms there are also many littoral or benthic species, quite common, that usually occur in clean and cold mountain waters. They are mainly oligohalobous (indifferent), pH circumneutral species.

On the other hand the subphase H2b (depth 967.5-957.5 cm) contains only some frustules of Fragilaria, Pinnularia, Eunotia. They usually occur in a littoral, mainly amidst epiphytes in bogs and springs, frequently also at wet rocks amidst mosses. This subphase cannot be easily defined as

these diatoms are very rare.

A renewed increase in number as well as in diatom taxa is noted in the subphase H2c (depth 955 cm). The genera Fragilaria, Pinnularia, Navicula, Diatoma, Amphora, Cymbella, Achnanthes are more numerous whereas Cyclotella, Asterionella, Eunotia, Diploneis, Neidium, Stauroneis, Anomeonis and Epithemia are less abundant. Such flora composition corresponds with the subphase H2a but in the subphase H2c there are more eu-terrestrial, atmophytic diatoms that are common amidst xerotic mosses and lichens on a wet ground and on rocks (Navicula mutica, N.contenta, N.purpusilla, Hantzschia amphioxys, Pinnularia borealis, Melosira roseana, Aulacoseira after R.M.CRAWFORD (1981). There is also a rheophile flora, particularly common in mountains, springs, streams, in clean cold flowing waters (Diatoma hiemale var. mesodon, Meridion circulare et var. constricta, Ceratoneis arcus).

2.3 Diatom phase H3 is represented (depth 952.5-645 cm) by not very abundant meso-

33

phytic and eurytopic diatoms, among which there are also some species of the genus Fragilaria. Besides there are still eu-terrestrial, atmophytic diatoms, living in dry or moist habitats (xerotic or meso-phytic mosses, dry or wet rocks, in a shallow water, and rheophile diatoms (Melosira) Aulacoseira roseana, Pinnularia borealis, Hantzschia amphioxys, Navicula contenta, N.perpusilla, N.mutica, Meridion circulare, Diatoma hiemale var. mesodon. Such diatom composition corresponds with the flora of the preceding phase (H2c). But the diatom phase H3 is different for a presence of many taxons that are typical for this phase (Navicula gibbula, Pinnularia molaris, P. lenticulata, Achnanthes coarctata, Stauro-neis parvula var. prominula, Cymbella si-nuata, Surirella angustata, Stauroneis smithii). The diatom phase H3 possesses oligohalobous (indifferent) diatoms, pH circumneutral that are most common in the mountain areas, as well as alkaliphile, eurytopic and eurytrophic ones that spread vastly and are mainly cosmopolitan. The presence of diatoms typical of varying eco-logic and hydrogeologic conditions sug-gests a non-uniform mosaic image of the environment in the present lake area. Pro-bably there were elements of a flora of flowing and stagnant waters, and of small depressions that could be occasionally fil-led with water. But this diatom develop-ment phase cannot be easily defined in time as it could occur in very short episodes, during which the flora changed its compo-sition very quickly, even seasonally.

2.4 Diatom phase H4 (depth 942.5-932.5 cm) is characterized by a sudden abundant de-velopment of a flora composed of Fragila-ria, mainly Fragilaria elliptica et f.mi-nor with a very high frequency (80-85%) in the whole phase.

Firstly during the subphase H4a, Fragi-laria elliptica et f.minor is accompanied by F.microstriata, F.brevistriata?, F.pseu-doconstruens et var. bigibba, Navicula wittrockii and to a smaller degree by Fra-gilaria alpestris. During the subphase H4b there are Fragilaria lapponica, Pinnularia borealis, Tabellaria flocculosa, Nitzschia sp., Navicula pseudoscutiformis. On the other hand an increase of Fragilaria pin-nata, F.alpestris, F.pseudoconstruens var. bigibba in the subphase H4c is significant.

Most of the diatoms found are species that have been previously very seldom or entirely not distinguished and have very poorly or unknown ecologic features.

Fragilaria elliptica was distinguished (in spite of its similarity to F.pinnata and F.construens var. venter) on the basis of comparative investigations in LM and SEM of very rich flora of Fragilaria in Late Glacial sediments of north-western Scotland (E.Y.HAWORTH 1975, 1976) as well as in Late Glacial and Holocene sediments of the Przedni Staw Lake, Tatra Mts. (B. MARCINIAK 1982, 1983).

A particular position in the diatom phase H4 is also occupied by Fragilaria ellipti-ca f. minor that is, after A.CLEVE-EULER (1953, p. 39) the plankton form, typical for eutrophic lakes in Lapland, living not only in a benthos but mainly in the photic zone. According to E.Y.HAWORTH (1976) this taxon is too small to be studied in the plankton and demands specific investiga-tions of the nannoplankton.

The described phase encloses also Fragi-laria pseudoconstruens et var. bigibba, considered to be a new species and a new variety as they are typical for Late Gla-cial and Holocene sediments of the high mountain Przedni Staw Lake in the Tatra Mts. (B.MARCINIAK 1982, 1983). Valves very similar to the ones of Fragilaria pseudo-construens were described by N.FOGED (1958, p. 68) from Greenland but they were ascri-bed by him to Fragilaria construens. On the other hand Fragilaria pseudoconstruens var. bigibba is very close or can be even iden-tical with Fragilaria robusta that occurs in a dysphotic zone of a volcanic Pavin Lake in France (E.MANGUIN 1954). According to Dr. M.K.HEIN of the USA (personel com-munication) Fragilaria pseudoconstruens et var. bigibba is noted now in a periphytic collection from Alaska.

Most species of Fragilaria live now in a littoral zone of water reservoirs, usually in stagnant and seldom in flowing waters. They are alkaliphile, vastly expanded and cosmopolitan species. The occurrence of Fragilaria pinnata (probably bryophilous) calls for attention as, according to N. FOGED (1958, 1972), it is noted in waters with pH circumneutral. It is common in western Greenland in alkalic sites at ba-salt outcrops and in neutral and slightly acid sites in gneiss areas. It reaches 99-100% in sediments of the Klarsø Lake in nothern Greenland. These taxonomic and ecologic data, although very incomplete, on some diatoms present in the diatom phase H4, do not allow to define the pa-laeohydrologic and palaeoecologic condi-tions of that time as they are often in-consistent with one another.

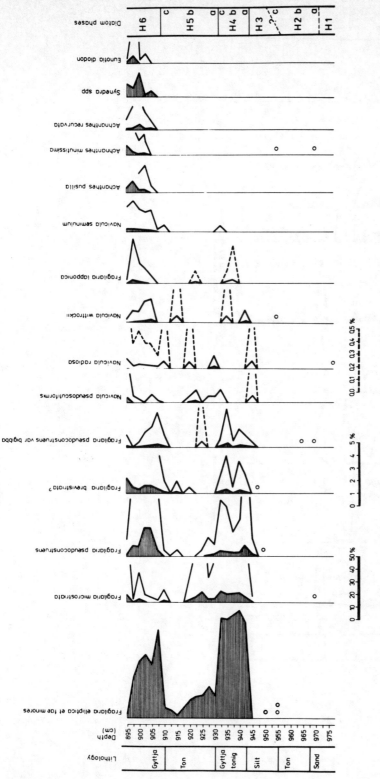

Figure 2a. Diatom diagram of core HO 2. Part I. O: Present in the sample.

35

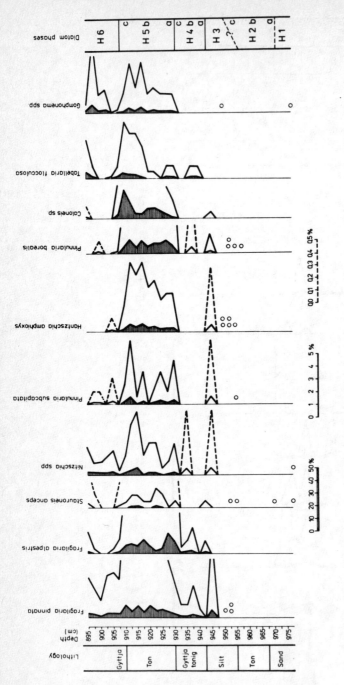

Figure 2b. Diatom diagram of core HO 2. Part II. O: Present in the sample.

36

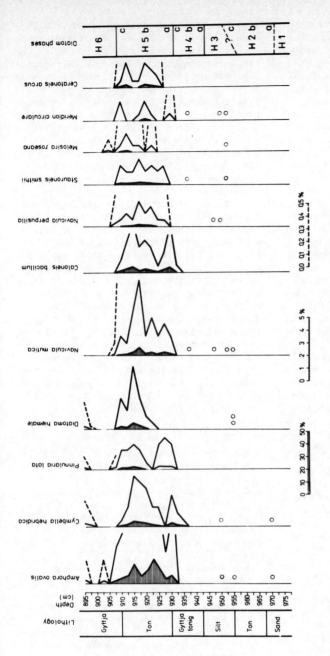

Figure 2c. Diatom diagram of core HO 2. Part III. O: Present in the sample.

2.5 Diatom phase H5 (depth 930-910 cm) indicates a considerable drop in a number of diatoms. Most species of the genus Fragilaria decrease (they predominated in the preceding phase) whereas the genera Tabellaria, Pinnularia, Hantzschia, Stauroneis, Diatoma, Ceratoneis, Melosira, Amphora, Navicula, Caloneis, Nitzschia are more important.

Firstly, in the subphase H5a Fragilaria elliptica et f.minor constitutes up to 25% of the diatom flora. F.microstriata is also quite abundant and there are more F. alpestris and F.pinnata, Caloneis, Navicula, Pinnularia, Cymbella hebridica, Hantzschia amphioxys, Nitzschia. In the subphase H5b a content of Fragilaria elliptica, F.microstriata and F.alpestris still decreases. There are quite numerous F.pinnata, Amphora ovalis, Ceratoneis arcus, Meridion circulare, Caloneis, Pinnularia. In the subphase H5c a development of Caloneis occurs, there are also Fragilaria pinnata, F.alpestris, Amphora ovalis, Pinnularia, Tabellaria, Nitzschia, Gomphonema, Diatoma, Ceratoneis, Cymbella, Melosira, Navicula.

In the diatom phase H5 there is a considerable participation of rheophile north-montane, cold-water and halophobous diatoms, noted in streams and springs of northern Europe and in mountains of Central Europe. There are also eu-terrestrial diatoms, common on dry or wet rocks and on mosses. Such composition of flora shows the contact of diatom communities of various habitats: flowing waters, streams, waterfalls as well as dry or only occasionally flooded shallow water reservoirs with influx or overflow. Some flora in such reservoir could have a local characteristic and contain lake diatoms (mesophytic or hydrophytic ones) whereas the other could be derived from the near surroundings (so allochthonous specimens brought into the reservoir by flowing water). This development phase of the Lake Hobschen is similar to the earlier subphase of the diatom phase H3 that also contained the diatoms typical for various habitats. The varying composition of flora can suggest quickly changing environment or can result from washing by water from the near surrounding of the diatoms, living in flowing waters, with eu-terrestrial and hydro-terrestrial specimens inclusive (sensu J.BOYE PETERSEN 1935).

2.6 During the diatom phase H6 (depth from 907.5 cm) a renewed mass development of some Fragilaria species occurred (F.elliptica et f. minor, F.pseudoconstruens, F. brevistriata?) accompanied by a smaller increase in contents of Navicula (N.radiosa, N.pseudoscutiformis, N.wittrockii, N.seminulum) and of the genus Achnanthes (A.pusilla, A.minutissima, A.recurvata) that, together with the genera Synedra and Eunotia as well as Fragilaria brevistriata?, is considerably more abundant during the second part of the described phase. In the first part of the diatom phase H6, a diatom flora (mainly Fragilaria and Navicula) corresponds highly with the succession of diatoms during previous diatom phases H2 and particularly H4. On the other hand in the second part there are diatoms that live in a pelagic zone of montane lakes and in waters of subarctic regions; there are also diatoms vastly expanded in stagnant and slowly flowing waters and springs of the whole Europe, but they are abundant in northern Europe and in mountain waters (smaller specimens are also noted on wet rocks and mosses). Diatoms that develop at lake bottoms and on lichens are common too; they are the cosmopolitan taxa. As no quantitative analysis was done for the higher part of the section, then it seems untimely to define the border of this phase and of the following one.

3 FINAL REMARKS

The preliminary analysis of diatoms from the lower part of the section of the Lake Hobschen proved considerable qualitative and quantitative variations of diatoms. Six phases of vegetative succession were distinguished, named the diatom phases (H1-H6). These phases are highly concordant with a lithologic subdivision of sediments (cf. Fig.2A).

A succession of diatoms in the analyzed sediments proves that the diatom phases H1, H2a, H4 and H6 contain mainly a littoral (mesophytic, hydrophytic) flora and smaller contents of hydroterrestrial elements (sensu J.BOYE PETERSEN 1935). This flora probably represents poorly developed limnic stages. A lake depth cannot be defined for the described phases. In extreme climatic conditions that can be compared with arctic ones, even in quite a deep lake that can thaw only partly in summer, there can be nothing more than a poor littoral flora and planktonic diatoms do not develop (H. BACHMANN 1928). There are also arctic lakes, the postglacial sediments of which do not contain the diatoms, for example

38

Lake Linnévatn in Spitsbergen (A.BØYUM & J.KJENSMO 1980).

On the other hand the diatom phases H2c, H3 and H5 represent the successive stages of a development of the eu-terrestrial flora, described among others by H.BEGER (1927), J.BOYE PETERSEN (1935), W.BOCK (1963). There are also rheophile and facultative terrestrial diatoms sensu M.B.FLORIN (1970). The eu-terrestrial diatoms are typical for a very dry (xerotic mosses) or moist (mesophytic mosses) environment and for dry and wet rocks. Such flora is very rare in ancient sediments. It has been preserved in the 1 cm thick series of Late Glacial sediments at Kirchner Marsh (south-eastern Minnesota) that constitute the first, pre-limnic stage of development of a supraglacial lake formed within a glacial lobe of the Wisconsin age during its disintegration into dead-ice blocks and depressions.

4 REFERENCES

BACHMANN,H., 1928: Das Phytoplankton der Pioraseen nebst einigen Beiträgen zur Kenntnis des Phytoplanktons schweizerischer Alpenseen. - Z.f.Hydrologie V, 1/2, H., 50-103.

BEGER,H., 1927: Beiträge zur Ökologie und Soziologie der luftlebigen (atmophytischen) Kieselalgen. - Ber. der Deutsch. Bot. Ges. XLV, 6, 385-407.

BOCK,W., 1963: Diatomeen extrem trockener Standorte. - Nova Hedwigia V, 199-254.

BØYUM,A. & J.KJENSMO, 1980: Post-Glacial sediments in Lake Linnévatn, Spitsbergen. - Arch. Hydrobiol. 88/2, 232-249.

CLEVE-EULER,A., 1953: Die Diatomeen von Schweden und Finnland II. - K. svenska Vetensk-Akad. Handl. Fjärde Ser.4/1, 158 pp.

CRAWFORD,R.M., 1981: The Diatom genus Aula-coseira Thwaites: its structure and taxonomy. - Phycologia 20/2, 174-192.

FLORIN,M.B., 1970: Late Glacial Diatoms of Kirchner Marsh Southeastern Minnesota. - Beich. z. Nova Hedwigia 31, 667-757.

FOGED,N., 1958: The Diatoms in the basalt area and adjoining areas of archean rock in West Greenland. - Meddel. om Grønland. 156/4, 146 pp.

FOGED,N., 1972: The Diatoms in four Postglacial deposits in Greenland. - Meddel. om Grønland 194/4, 66 pp.

HAWORTH,E.Y., 1975: A scanning electron microscope study of some different frustule forms of the genus Fragilaria found in Scottish late-glacial sediments. - Br. phycol., J. 10, 73.

HAWORTH,E.Y., 1976: Two late-glacial (Late Devensian) diatom assemblage profiles from Northern Scotland. - New Phytol. 77, 227-256.

MANGUIN,E., 1954: Contribution à la connaissance des boues lacustres Lac Pavin(Puy de Dome). - Ann. Ec. Nat. Eaux et Forêts et de la Station de Recherches et experiences 14/1, 5-19.

MARCINIAK,B., 1982: Late Glacial and Holocene new diatoms from a glacial lake Przedni Staw in the Pięć Stawów Polskich Valley, Polish Tatra Mts. - Acta Geol. Hungar. 25, 1/2, 161-171.

MARCINIAK,B. & A.CIEŚLA, 1983: Diatomological and geochemical studies on Late Glacial and Holocene sediments from the Przedni Staw Lake in the Dolina Pięciu Stawów Valley (Tatra Mtes.). - Kwart. Geolog. 27/1, 123-150.

MILLER,U., 1982: Weichselian stratigraphy in South-Western Scania Sweden, according to microfossils. - Quaternary Studies in Poland 3, 69-77.

PETERSEN, J.BOYE, 1935: Studies on the Biology and Taxonomy of soil Algae. - Dansk. Botanisk Arkiv 8/9, 140-183.

Lake, Mire and River Environments, Lang & Schlüchter (eds)
© *1988 Balkema, Rotterdam. ISBN 90 6191 849 9*

Palynological and isotope studies on carbonate sediments from some Polish lakes – Preliminary results

K.Rozanski & D.Wcisło
Institute of Physics and Nuclear Techniques, University of Mining and Metallurgy, Cracow, Poland

K.Harmata
Botanical Institute, Jagellonian University, Cracow, Poland

B.Noryskiewicz
Institute of Geography, M.Copernicus University, Toruń, Poland

M.Ralska-Jasiewiczowa
Institute of Botany, Polish Academy of Sciences, Cracow, Poland

ABSTRACT: Presented are preliminary results of palynological and isotope ($\delta^{18}O$, $\delta^{13}C$) investigations performed on lake sediment cores originating from three sites in Poland: a fossil oxbow lake of late-glacial age (Roztoki, SE Poland), lake Strażym situated near Brodnica (NC Poland) and lake Mikołajki belonging to the East Baltic Lake Districts (NE Poland). In general, good agreement was found between palynological and isotope data. Periods of colder climate as indicated by pollen diagrams, are also seen on the $\delta^{18}O$ profiles as a distinct minima in the ^{18}O content of the lake carbonate. The $\delta^{13}C$ data are discussed in terms of possible changes in biological activity of the investigated lakes and/or climate induced fluctuations in their water level. The rise of temperature at the beginning of Holocene as inferred from the N-Polish isotope profiles, with its first culmination occuring close to the end of Preboreal, seems to have been delayed by several hundreds of years in comparison with the south-west areas of central Europe.

1 INTRODUCTION

As demonstrated by STUVIER (1970), the ^{18}O method of reconstructing the changes of palaeotemperature, which was tested with great success in marine environments, can also be applied in the palaeoclimatic studies of freshwater carbonates. In recent years several interesting examples of combined pollen and isotope studies on lacustrine sediments from Europe and Canada have been published (EICHER et al. 1981, EICHER & SIEGENTHALER 1976, 1983, MÖRNER & WALLIN 1977, FRITZ et al. 1975, LEMEILLE et al. 1983, FRITZ 1983, PUNNING 1984). The paper presents preliminary results of oxygen and carbon isotope investigations carried out on carbonate sediment cores taken out from some Polish lakes. The results are discussed in connection with other data available for these cores ($CaCO_3$ content, pollen analyses, ^{14}C-dates)

The oxygen isotopic composition of the lake carbonate is controlled by the $^{18}O/^{16}O$ ratios of the lake water. The $^{18}O/^{16}O$ ratios of water, bicarbonate and carbonate, are related by the known temperature-dependent isotopic fractionation factors, assuming that the thermodynamic equilibrium is maintained during the precipitation of carbonate. If the isotopic composition of water and carbonate is expressed in δ-notation, where δ is defined as a permille deviation from the internationally accepted standards (SMOW for water samples {CRAIG 1961} and PDB for carbonates {CRAIG 1957}), then the lake temperature T (in degrees Celsius) and the isotopic difference $\delta^{18}O_{carb}$ - $\delta^{18}O_{water}$ are related by the following equation (CRAIG 1965):

$$T=16.1-4.21 \cdot (\delta^{18}O_c - \delta^{18}O_w) + 0.13 \cdot (\delta^{18}O_c - \delta^{18}O_w)^2 \tag{1}$$

Knowing both $\delta^{18}O_c$ and $\delta^{18}O_w$ one can, in principle, calculate the temperature of the lakewater characteristic for the period of carbonate precipitation. Because the precipitation of carbonates occurs mainly during summer months when the lake is biologically active, the temperature in eq.(1) will reflect the mean summer temperature of the lake water.

The ^{18}O content of the lake water in the past cannot be directly determined. Therefore, only a very rough assessment of the lake temperature is possible on the basis of eq.(1). However, instead of estimating the absolute temperature from eq.(1), we can consider the temperature dependence on

$\delta^{18}O_c$, and judge the changes in the lake temperature from variations of the ^{18}O content in the lake carbonate along the core.

The change of $\delta^{18}O_c$ with changing temperature can be approximated by the following equation:

$$\delta^{18}O_c=(-0.24‰/°C+A+B)\cdot\Delta T \qquad (2)$$

where:

A – apparent isotope (temperature coefficient for the $\delta^{18}O$ variations in precipitation. We assume here the range between 0.37 and 0.58‰/°C (ROZANSKI) 1984).

B – evaporative enrichment of the lake water in ^{18}O per degree Celsius. This coefficient was estimated to be between 0.04 and 0.08‰/°C (WCISŁO 1985).

Inserting the above values for A and B into eq.(2) we shall get the following:

$$\delta^{18}O_c=(0.17÷0÷42)\cdot\Delta T \qquad (3)$$

This equation does not take into account such effects as a change in the isotopic composition of the ocean induced by the preferential storage of light isotopes in the continental ice-sheets or the presumable variations in lake's characteristics (depth, mean residence time of water, relative humidity above the lake surface, etc.).

It becomes evident from the above discussion that the conversion of the apparent variations in ^{18}O content of the lake carbonates into the changes in the surface air temperature (via the lake temperature) is biased by relatively high uncertainty.

The interpretation of the $^{13}C/^{12}C$ ratios in the lacustrine carbonates is even more complex than in the case of the $^{18}O/^{16}O$ ratios. The $\delta^{13}C$ valus of the lake carbonate are controlled by many factors, the most important among them being:

- ^{13}C content of the bicarbonates dissolved in the water feeding the lake,
- degree of exchange between the lake water and the atmospheric CO_2,
- stratification of the lake,
- biological activity of the lake.

The biological activity is often considered as a decisive factor; due to large kinetic isotope effect during the assimilation process in the lake, the remaining carbonate ions are enriched in ^{13}C. Because at mid latitudes the shallow groundwaters and rivers have in general $\delta^{13}C$ between -15 and -10‰, a complete equilibration of the lake water with atmospheric

CO_2 ($\delta^{13}C \simeq -7‰$) in the open system conditions would lead to $\delta^{13}C$ of the lake carbonates close to + 2‰, if other sources of carbon are insignificant.

2 ANALYTICAL PROCEDURE

The $\delta^{18}O$ and $\delta^{13}C$ of the lake carbonate samples were measured on the mass spectrometer Micromass 602C (Vacuum Generators Ltd.). The conversion of samples into CO_2 was done using the standard technique developed for marine carbonates (McCREA 1950). Additionally, the amount of CO_2 was measured, thus providing the information on the $CaCO_3$ content in the sample. No special treatment of samples except of drying ($\sim$110°C, 10 hours) was used. The details of the preparation and analytical procedures are discussed elsewhere (ROZANSKI & WCISŁO 1985).

3 THE INVESTIGATED MATERIAL

The isotope analyses have been performed on three sediment cores originating from the following sites:

A. Roztoki, 250 m a.s.l., 49.75°N, 21.55°E, SE Poland

The site is located on the terrace of the Jasiołka River (4 km east from Jasło) in the intermontane Jasło-Sanok Depression, in West Carpathians (Fig.1). It is a fossil oxbow lake of late-glacial age that had been overgrown and then filled with peat in connection with the lowering of water level at the beginning of the Holocene. The lake was subsequently covered by flood deposits interrupting the peat growth. The profile was collected in 1982, from an exposure in the central part of the lake. The results of pollen analysis carried out by K.HARMATA (ALEXANDROWICZ et al. 1985) are presented in Fig. 2 in a very simplified form to show only the main features of the vegetational development.

B. Lake Strażym, 71 m a.s.l., 53.18°N, 19.26°E, NC Poland

The lake situated 10 km north from Brodnica belongs to the Brodnickie Lake District, one of the easternmost regions of the South Baltic Lake Districts (KONDRACKI 1978) - Fig. 1. The young morainic lanscape of this region was formed by the Vistulian glaciation. The recent eutrophic lake covers the area of 0.73 km^2 and has a mean depth of ca. 3 m, but its sediments provide an evidence of strong water-level oscilla-

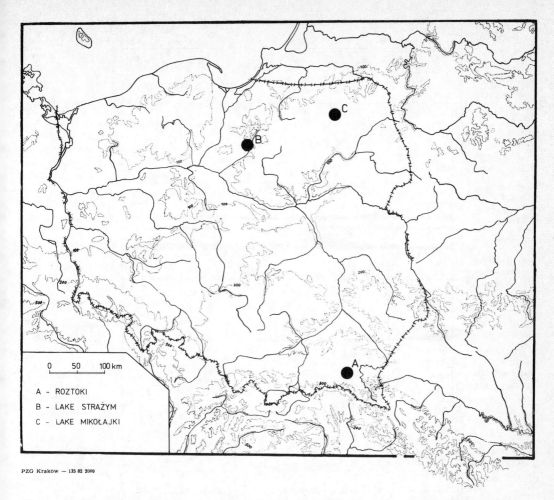

Figure 1. The map showing the location of investigated sites. A - Roztoki fossil lake,
B - Lake Strażym, C - Lake Mikołajki.

0   50   100 km

A - ROZTOKI
B - LAKE STRAŻYM
C - LAKE MIKOŁAJKI

tions and changes of the lake extent throughout its history. The core was taken in the marginal southern part of the lake in 1984, by means of the Livingstone type corer (WIECKOWSKI 1970).

C. Lake Mikołajki, 116 m a.s.l., 53.46°N, 21.35 °E, NE Poland

The lake Mikołajki lies in the Masurian Lake District, within the region of East Baltic Lake Districts belonging to the Lowlands of Eastern Europe(KONDRACKI 1978). It forms the middle part of the lake channel (38 km long, subglacial origin), connected with the system of the biggest lacustrine basin of Poland - Lake Śniardwy.

The Lake Mikołajki itself is 5.4 km long, 4.6 km^2 in area, with the mean depth of 11.0 m. It is a typical eutrophic lake. The mean residence time of water in the lake has been estimated at around 4 months, with the mean summer temperature of the lake equal to 13.6°C (PIECZYŃSKA 1976). The palaeobotanical studies of sediment cores from the central part of the lake were carried out by M.RALSKA-JASIEWICZOWA (1966). The new core was taken out in 1984 in the northern part of the lake by the Livingstone type piston corer (WIECKOWSKI 1970).

43

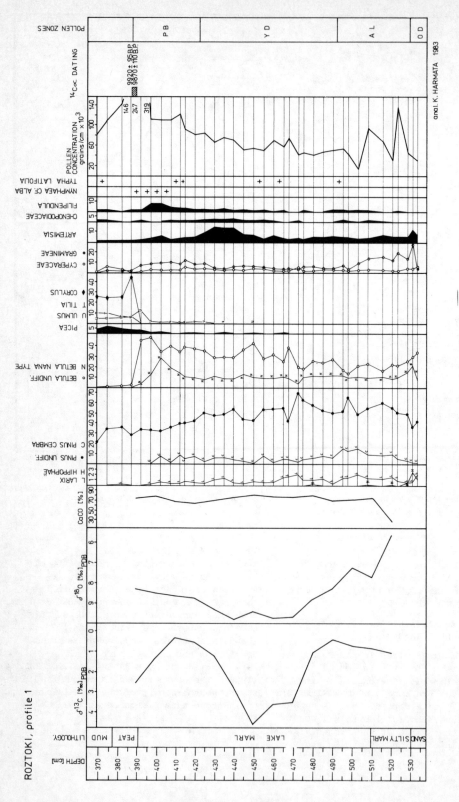

Figure 2. Isotope and pollen diagram from Roztoki fossil lake (Profile 1).

anal. K. HARMATA 1983

4 DISCUSSION

Roztoki

The marl sediments in the lake were accumulated during the time span from the end of the Older Dryas till the beginning of the Holocene. In the pollen diagram covering this part of the profile several phases of vegetational development have been recorded (ALEXANDROWICZ et al. 1985).

A short phase with the partly open vegetation dominated by Gramineae-Artemisia assemblages, with shrubs of Betula nana and Hippophaë and groups of trees (Pinus sylvestis, Larix and Betula) recorded in the basal sand up to 530 cm, corresponds most probably with the end of the Older Dryas. It is followed by a zone showing the development of coniferous forests with the high contribution of Pinus cembra and Larix, typical for Allerød time (530-495 cm). In the next zone (494-450 cm) the birchwoods spread gradually, the role of conifers being more and more reduced. It is not quite clear what time exactly this phase reflects.

The expansion of open communities with Artemisia and Chenopodiaceae in the following zone (450-435 cm) is an obvious reflection of the Younger Dryas climatic oscillation. The beginning of the Holocene is pronounced by a successive spread of birchwoods and dwarf shrubs (430-395 cm). The rapid change óf pollen spectra around 390 cm showing the development of deciduous forests with abundant hazel accompanied by the spread of spruce, corresponds with the sharp change of sediment type from the lake marl to peat. The beginning of peat deposit has been dated with two ^{14}C-dates: 9920 ± 95 BP and 9879 ± 110 BP. Accumulation of peat is connected with a strong reduction of deposition rates, and with rise in pollen concentration.

The $\delta^{18}O$ curve exhibits a rather regular character, with the maximum values in its bottom part corresponding to the Allerød period, and a broad minimum at the depth of 400-470 cm synchronous with the birchwood phase and Artemisia phase of the Younger Dryas. It is followed by a gradual increase of the ^{18}O content at the transition to the Holocene, up to -8.2‰ in the uppermost sample of the lake marl. If we adopt the estimated upper limit of the $\delta^{18}O$/temperature coefficient (eq.3), then the change in the ^{18}O content between 500 and 470 cm would suggest the diminishing of the mean summer temperatures by about $5^{o}C$ which is surprisingly much as for the

Allerød/Younger Dryas transition. The pollen diagrams do not show correspondingly drastic changes in the vegetation. The subsequent improvement of climate, as suggested by the upper part of the $\delta^{18}O$ curve, would be accompanied by the increase of summer temperature of ca. $3^{o}C$. The relatively small number of experimental points in the isotope profile does not allow at present any more detailed interpretation.

The $\delta^{13}C$ curve shows also a distinct minimum in the central part of the profile, that might be explained by the reduced biological activity of the lake due to cooler climate. The high $\delta^{13}C$ values in the bottom and upper parts of the marl deposit could be connected with the shallow environment, when the carbon budget in the lake was controlled by the intensive exchange with the atmospheric CO_2.

Lake Strazym

The accumulation of calcareous gyttja sediment in the investigated part of the lake started in the Preboreal period. It terminated around 6000 BP by the lowering of the water level which further resulted in the overgrowing of this part of the lake by a reedswamp. During the subsequent rise of water level the top layers swamp peat have been destroyed. The uppermost existing part of this peat has been dated at 3960 ± 120 BP whereas its bottom part has the ^{14}C-age of 5890 ± 140 BP. The simplified pollen diagram in Fig.3 is based on unpublished data obtained by B.NORYŚKIEWICZ.

The vegetational development recorded in this part of profile starts with a short birchwood phase (490-460 cm) typical for the early Preboreal, followed by the spread of pine forests with sporadically appearing Ulmus, and later also Corylus (460-410 cm , middle late Preboreal). The first spread of Corylus opening the Boreal period can be dated in this area at ca. 9200 BP (RALSKA-JASIEWICZOWA 1983). The Atlantic phase was characterized by some development of deciduous forests with Quercus, Ulmus and a little Tilia, still with the pine-forests prevailing all the time.

The $\delta^{18}O$ curve starts with a distinct minimum between 470 and 410 cm, corresponding with the Preboreal time (the lowermost sample is not representative due to its low $CaCO_3$ content). The rapid rise of $\delta^{18}O$ between 410 and 390 cm is synchronous with the increase in Corylus pollen frequencies and thus can approximately be referred to the time just before 9200 BP. In

LAKE STRAŻYM, profile 1

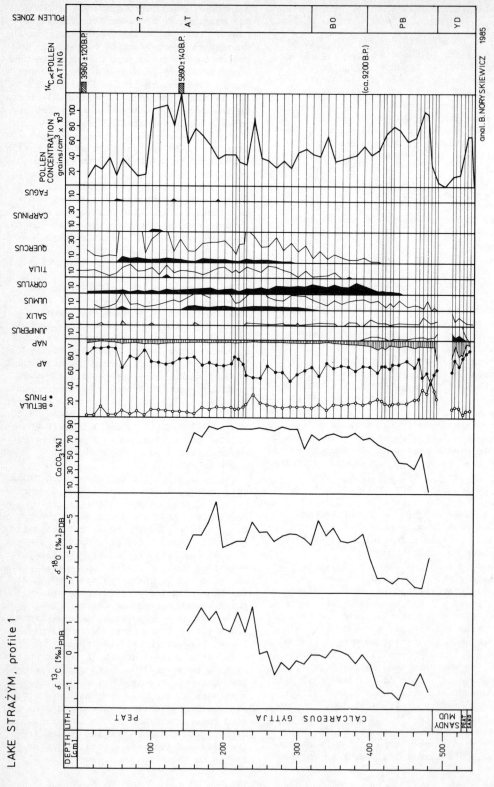

Figure 3. Isotope and pollen diagram from the Lake Strażym.

46

the overlying sequence of the sediment $\delta^{18}O$ values remain high, with some superimposed minor fluctuations.

The $\delta^{13}C$ curve parallels in general that of ^{18}O content, especially in its bottom and middle parts. Between 260 and 240 cm a rapid rise of ^{13}C content by about 2‰ is observed. Age of this change may be calculated roughly at ca. 7300-7000 BP. Since the upper part of the core represents a stage of lake overgrowing (150 cm of peat), it seems reasonable to explain this increase by the progressing reduction of water depth at the lake margin, leading to intensive exchange between the lake carbonates and atmospheric CO_2.

Around 6000 BP the accumulation of calcareous gyttja was stopped by the lowering of water level, and the lake margin was overgrown with a reedswamp.

Lake Mikołajki

The investigated profile, 550 cm deep, covers the late Glacial starting with the decline of the Older Dryas (12 000 ± 130 BP) and continues throughout the whole Holocene. During the late Glacial mostly telmatic sediments were deposited at the coring place, indicating the shallow water with progressing overgrowing of the lake. The maximum lowering of water level was reached during the early Preboreal, when the peat composed mostly of Dryopteris thelypteris was deposited. It was followed by a gradual rise of water, and in the late Preboreal the carbonate sediments started to accumulate. The ^{14}C-date just below this change in type of sediment at 365 cm is 9450 ± 100 BP. The $CaCO_3$ content in the calcareous gyttjas ranges from 60 to 80% during the most part of Holocene to ca. 30 + 50% in the last centuries.

Until now only the bottom part of the profile up to 290 cm has been regularly investigated by means of pollen analysis (M.RALSKA-JASIEWICZOWA - unpublished results) revealing the Older Dryas, Allerød, Younger Dryas and early Preboreal vegetational development recorded in peat and peaty gyttja deposits. The time around 9600 BP, when the carbonate sediments started to accumulate in the investigated part of the lake, corresponds to the transition from the early Preboreal zone of pine-birch forests to the late Preboreal zone with similar type of forest communities dominating, but with some contribution of elm and hazel. The first distinct rise in Corylus pollen curve has been recorded in the diagram at 290 cm, and its

age may be estimated as ca. 9000 BP. For the upper part of the core only the pilot pollen counts have been performed around the levels of more important changes in the isotope curve, and their approximate age was calculated by comparison with the pollen diagram from the central segment of the lake channel(RALSKA-JASIEWICZOWA 1966).

The $\delta^{18}O$ curve exhibits some oscillations in its bottom part, with the minimum ^{18}O content (-8.9‰) in the lowermost sample at 365 cm dated at ca. 9450 BP, another depression (-8.5‰) in the laminated part of the core around 325 cm, and a small peak (-8.0‰) inbetween. However it generally tends towards a rise, culminating with the value of ca. -7‰ at the level of 300 cm, just beneath the first rise in Corylus pollen curve for which the assumed age is around 9000 BP. Between 300 and 170 cm the $\delta^{18}O$ curve shows tendency towards slight rise, followed by a decreasing trend with some minor fluctuations since 170 cm and a more distinct decline close to the sediment surface. The age calculation for 170 cm level derived from the correlation of pollen counts at and around this level with the pollen diagram published in 1966 (RALSKA-JASIEWICZOWA) yields the value around 5500 BP. This estimate has been based primarily on the position of 170 cm level against the level of first Ulmus fall and on the presence of single Plantago lanceolata pollen grains.

Accepting the estimated upper limit for the $\delta^{18}O$/temperature coefficient (eq.3), the increase in the ^{18}O content between ca. 9500 and ca. 9100(9200)BP would correspond to the change in the mean summer temperature of the lake of ca. 4 - 5°C. It is remarkable that this shift in $\delta^{18}O$ correlates well with the shift of similar magnitude (ca. 1.5‰) observed in the Strazym Lake, and occurred at a similar time. The increase of the mean summer temperature suggested by the ^{18}O data between ca. 9000 and 5500 BP would be about 1°C, whereas the subsequent slight cooling up to the top of the core would be about 1.0 to 1.5°C.

The profile of ^{13}C content shows very high values in its bottom part with the maximum of +1.0‰ at 360 cm, declining rapidly to -2.5‰ at 330 cm. This elevated ^{13}C content suggests the shallow water and/ or the high biological activity of the lake at the time around 9400 BP and is in agreement with the results of both pollen and sediment analyses indicating the gradual rise of water level just after the

LAKE MIKOŁAJKI / 1984

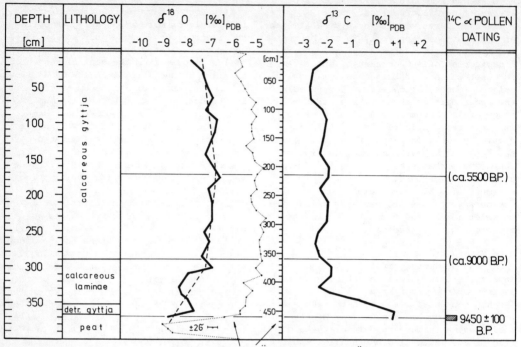

DEPTH [cm]	LITHOLOGY	δ¹⁸O [‰]$_{PDB}$	δ¹³C [‰]$_{PDB}$	¹⁴C ∝ POLLEN DATING

LAKE TINGSTÄDE TRASK, GOTLAND (MÖRNER, WALLIN 1977)

Figure 4. $\delta^{18}O$ and $\delta^{13}C$ curves from Lake Mikołajki core 1984. The broken curve indicates general trends in changes of the $\delta^{18}O$ profile.

shallowing phase in the early Preboreal. The subsequent decline of $\delta^{13}C$ was a consequence of this rise of water level. The remaining upper part of $\delta^{13}C$ curve parallels in general that of the ^{18}O content.

In the Fig.4 the $\delta^{18}O$ curve obtained from the Lake Tingstade Trask, Gotland (Sweden) and published by MÖRNER and WALLIN (1977) has been shown for comparison. This curve, although shifted by about 2‰ towards more positive values, reveals striking similarities to the $\delta^{18}O$ profile from the Mikołajki Lake in spite of small differences in their fine structure. These similarities speak for the consistence of major climatic trends during the Holocene over the large areas of Europe.

The rise of temperature at the beginning of the Holocene as inferred from the N-Polish profiles, with its first culmination occuring close to the end of Preboreal, seems to have been delayed by several hundreds of years in comparison with the southwest areas of central Europe. This effect in $\delta^{18}O$ may, however, be connected with the

still unsettled sedimentation processes during early stages of lakes formation. The first Holocene peak of $\delta^{18}O$ curve in central-European profiles is most often pronounced in the early Preboreal time (EICHER & SIEGENTHALER 1976, 1983; EICHER et al. 1981).

5 REFERENCES

ALEXANDROWICZ,S.W., K.HARMATA & A.WÓJCIK, 1985: Sedimentation of lacustrine and fluvial deposits in the Jasiołka valley. Genesis of ancient outletless hollows and stratigraphy of filling them up with sediments. – Carpatho-Balkan Geological Association XIII Congress. Cracow, Poland. Guide to excursion 5, 79-82.

CRAIG,H., 1957: Isotopic standards for carbon and oxygen and correction factors for mass-spectrometric analysis of carbon dioxide. – Geochim.Cosmochim.Acta 12, 133.

CRAIG,H., 1961: Standard for reporting concentrations of deuterium and oxygen-18 in natural waters. – Science 133, 1833.

CRAIG,H., 1965: The measurement of oxygen isotope paleotemperatures. Stable Isotopes in Oceanographic Studies and Paleotemperatures. - Spoleto 1965, 161-182.

EICHER,U. & U.SIEGENTHALER, 1976: Palynological and oxygen isotope investigations on late-Glacial sediment cores from Swiss lakes. - Boreas 5, 109-117.

EICHER,U. & U.SIEGENTHALER, 1983: Stable isotopes of oxygen and carbon in the carbonate sediments of lac de Siguret (Hautes-Alpes,France). - Ecologia Mediterranea 9, 49-53.

EICHER,U., U.SIEGENTHALER & S.WEGMÜLLER, 1981: Pollen and oxygen isotope analyses on late- and post-Glacial sediments of the Tourbière de Chirens (Dauphiné, France). - Quat. Res. 15, 160-170.

FRITZ,P. T.W.ANDERSON & C.F.M.LEWIS, 1975: Late-Quaternary climatic trends and history of lake Erie from stable isotope studies. - Science 190, 267-269.

FRITZ,P., 1983: Palaeoclimatic studies using freshwater deposits and fossil groundwater in central and northern Canada. In: Palaeoclimates and Palaeowaters: A Collection of Environmental Isotope Studies. - IAEA,Vienna, 157-166.

KONDRACKI,K., 1978: Geografia fizyczna Polski (Physical Geography of Poland), PWN, Warszawa 1978.

LEMEILLE,E., R.LETOLLE, F.MELIERE & P.OLIVE, 1983: Isotope and other physico-chemical parameters of palaeolake carbonates. In: Palaeoclimates and Palaeowaters: A Collection of Environmental Isotope Studies. IAEA, Vienna, 135-151.

McCREA,J.M., 1950: On the isotopic chemistry of carbonates and a paleotemperature scale. - J. Chem. Phys. 18, 849-853.

MÖRNER,N.A. & B.WALLIN, 1977: A 10 0000-year temperature record from Gotland, Sweden. - Palaeog. Palaeoclim. Palaeoecol. 21, 113-138.

PIECZYNSKA,E. (ed.) 1976: Selected Problems of Lake Littoral Ecology. - Univ. of Warsaw Press.

PUNNING, J.M., T.MATRA & R.VAIKMAE, 1984: Light isotope variations in carbonate sediments and their palaeogeographical value. - ZfI-Mitteilungen 84, 329-337.

RALSKA-JASIEWICZOWA,M., 1966: Osady denne Jeziora Mikołajskiego na Pojezierzu Mazurskim w świetle badań paleobotanicznych. (Bottom Sediments of the Mikołajki Lake - Masurian Lake District - in the Light of Palaeobotanical Investigations). - Acta Palaeobotanica 7, (2).

RALSKA-JASIEWICZOWA,M., 1983: Isopollen maps for Poland 0-11,000 BP. - New Phytol. 94, 133-175.

ROZANSKI,K., 1984: Temporal and spatial variations of deuterium and oxygen-18 in European precipitation and groundwaters. - ZfI-Mitteilungen 84, 341-353.

ROZANSKI,K. & D.WCISŁO, 1985: Oxygen and carbon isotope investigations of carbonate sediment cores from some Polish lakes. - Isotopenkolloquium, 5.-11. Sept. 1985, Freiberg, GDR.

STUVIER,M., 1970: Oxygen and carbon isotope ratios of freshwater carbonates as climatic indicators. - J. Geophys. Res. 75, 5247-5257.

WCISŁO,D., 1985: ^{18}O and ^{13}C isotopes in palaeoclimatic studies of lacustrine sediments. - Master Thesis Inst. of Physics and Nuclear Tech., Cracow 1985, (in Polish).

WIĘCKOWSKI,K., 1970: New type of lightweight piston core sampler. - Bull. Acad. Pol. Sc. Ser. Sc. Geol. Geogr. 18(1), 57-62.

Lake, Mire and River Environments, Lang & Schlüchter (eds)
© 1988 Balkema, Rotterdam. ISBN 90 6191 849 9

Isopollen maps of *Picea abies*, *Fagus sylvatica* and *Abies alba* in Czechoslovakia – Their application and limitations

E.Rybníčková & K.Rybníček
Institute of Experimental Phytotechnology, Czechoslovak Academy of Sciences, Brno, Czechoslovakia

ABSTRACT: Isopollen maps of Picea abies, Fagus sylvatica and Abies alba for each millenium after 12 000 BP are commented upon as examples of progressing pollen mapping in Czechoslovakia. In addition to the main reasons for constructing pollen maps, such as information on the distribution and migration routes of plants in the past, pollen mapping can also contribute to the determination of historical sources of present tree populations or their small taxa. Isopollen maps can also help to reconstruct past hydrological situation using pollen producers as indicators.

Among the three main climax trees mentioned above, Picea abies is supposed to have survived the last glaciation in several isolated spots in this country; two other trees are supposed to be immigrants after 7000 BP (Fagus) and 5000 BP (Abies). The level of distribution of all three trees indicates well the grade of air humidity of the upper areas (uplands, mountains) or high local humidity (waterlogging) of soils at middle and low altitudes.

1 INTRODUCTION

Isopollen mapping offers a substantial contribution to our knowledge of the past distribution of trees and shrubs or other plants as well as their migration routes, and may also help in the location of probable refugia, though the distribution of pollen need not reflect the actual distribution of pollen producers. Some papers, concerning various regions or countries, containing isopollen maps have already been published (cf. BIRKS 1985) and, recently, an atlas of isopollen maps for the whole Europe was issued (HUNTLEY & BIRKS 1980). While this atlas gives a general idea of tree or other pollen distributions since 13 000 BP, the regional isopollen maps, which present more details, can suggest answers to some more particular questions, such as explaining the historical sources of present tree populations or small taxa (varieties, forms, etc.) in a given region. Regional pollen mapping can also be used for indirect determination of regional hydrologic or climatic conditions in the past, using pollen producers as indicators of both air and soil humidity.

All these reasons have induced us to construct isopollen maps for the main forest trees in Czechoslovakia. That country has a very important position in Europe, being situated between the Alps and the Carpathians. In the past it represented a typical periglacial area between the Continental and Alpine ice sheets; and currently it has higly variable climatic, geological, geomorphic and soil environments. It can be assumed with a high degree of certainty that the flora and vegetation reacted to these conditions.

2 METHODS

In constructing isopollen maps we were in a highly favourable position, for we had at our disposal maps or information from nearly all neighbouring countries: Poland (RALSKA-JASIEWICZOWA 1980), West Germany (isochrones by BEUG 1981, MS), Austria (KRAL 1979), and the above-mentioned general outline for all of Europe (HUNTLEY & BIRKS, op. cit.). We used the same approach, method and recalculation produced in Poland by SZAFER (1935) and recently modernized and adapted by RALSKA-JASIEWICZOWA (op. cit.). As the basis of our work we used 64 sites with 75 pollen diagrams from

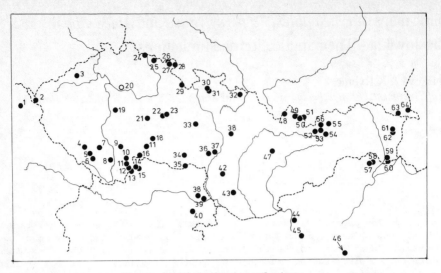

Figure 1. Distribution of sites.

Czechoslovakia and neighbouring areas across the borders (Tab.1 and Fig.l). Only four of these diagrams were produced before 1950, all the others are younger. Twenty-two of them have not been published yet; we would like to express our thanks to our colleagues Hüttemann, Jankovská, Konětopský, Krippel, Peichlová, Rypl and Svobodová, who put their original material at our disposal. Twenty-four profiles were dated by C[14], the others were synchronized with them. We did not always respect chronozones drawn by the authors of the diagrams. From each region we tried to select the diagrams that covered the longest period of time and seemed to reflect the regional vegetation. Whenever possible, we respected the isopollen lines drawn by RALSKA - JASIEWICZOWA (1980) along the Polish border. Where the data from our sites did not correspond to the Polish conclusions, we recommend a compromising line. In case the basic data allowed two or more possibilities of drawing the line, we chose the one that respects the expected geomorphic and climatic barriers.

Although the network of pollen diagrams used for our isopollen maps is comparatively dense, some "white areas" still exist, such as the Šumava region, Central Bohemia, Slovenské Rudohorie Mts, and some others. Our isopollen lines are very uncertain in these parts. In other areas the isopollen maps present a picture of the contemporary state of knowledge and, with an increased number of analyses, they will certainly undergo a more or less pronoun-

ced change. The ideal situation when the isopollen maps will not change after new pollen data have been produced is too distand in the future - in this country at least.

3 RESULTS

For this presentation we have chosen three main climax forest trees, viz. Picea abies, Fagus sylvatica and Abies alba. Although both conifers are retreating as a result of air and soil contamination and, secondarily, due to their higher sensitivity to animal and fungus pests, they still rank among the most important (Picea) or important (Abies) economic trees. Historical studies could contribute to analyses of the present populations and to the selection of best native forms in the future.

As far as the possibility of using pollen producers as indicators of humidity is concerned, the following should be said: Picea indicates areas with high humidity of both air and soil, and at the middle altitudes sometimes local waterlogging of soils. Fagus sylvatica requires, above all, high air humidity; it does not like wet or waterlogged sites with late ground frosts. Abies alba is very sensitive to low air humidity, and it does not like great variation in temperature and humidity; this means that it prefers northern slopes of mountain chains or valleys and wet but not waterlogged soils.

The isopollen maps for Picea cover every millenium from 11 000 (11 500) yrs BP to

Table 1. List of sites.

Nr.	Site	C 14	10^3 ys.BP													Author, year
			0,5	1	2	3	4	5	6	7	8	9	10	11	12	
1	Seelohe		+	+	+	+		+	+	+	+	+	+	+	+	Firbas 1937; Firbas 1949
2	Frant.Lázně		+	+	+	+	+	+	+	+	+					Pacltová 1972
3	Komořany	4	+	+	+	+		+	+	+	+	+	+	+	+	Jankovská MS; Losert 1940
4	Němčice		+	+	+											Rybníčková 1973
5	Nahořany		+	+	+											Rybníčková 1973
6	Lštění		+	+	+	+	+									Rybníčková 1964
7	Řežabinec	9	+	+	+	+	+	+			+	+	+			Rybníčková et Rybníček 1985
8	Zbudovská blata 1,2	4	+	+	+	+	+	+			+	+	+	+		Rybníčková et al. 1975
9	Borkovice 1,2	4	+	+	+	+	+	+	+	+	+	+	+	+	+	Jankovská 1980
10	Mokré louky	5	+	+	+	+	+	+	+	+	+	+	+			Jankovská MS
11	Branná		+	+	+	+	+	+	+							Jankovská 1980
12	Červené blato		+	+	+	+		+	+	+			+	+	+	Jankovská 1980
13	Velanská cesta		+	+							+	+	+	+	+	Jankovská 1980
14	Kiensass	4	+	+	+	+	+	+	+	+	+	+	+	+	+	Peschke 1977
15	Haslau	5	+	+	+	+	+	+	+	+	+	+	+	+		Peschke 1977
16	Bláto 1,2	2	+	+	+	+	+	+	+	+	+	+	+	+		Rybníčková 1974
17	Kásná	1	+	+	+	+	+	+	+	+	+	+	+			Rybníčková 1974
18	Loučky	1	+	+	+	+					+	+	+			Rybníčková 1974
19	Markvart		+	+	+											Pacltová 1955
(20)	Lysá-Hrabanov		Low representation of tree pollen has been considered only													Losert 1940
21	Palašiny		+	+	+	+	+	+	+	+	+	+	+			Jankovská MS
22	Zalíbené	2								+	+	+	+			Kneblová-Vodičková 1966a
23	Kameničky	10	+	+	+	+	+	+	+	+	+	+	+	+		Rybníčková MS
24	Pančická louka	6	+	+	+	+	+	+	+	+						Hüttemann 1983, MS
25	Černá hora		+	+	+	+	+									Pacltová 1957
26	Vernéřovice	9	+	+	+	+	+	+	+	+	+	+	+			Peichlová 1978, MS
27	Březová		+	+	+	+	+	+	+	+						Peichlová 1978, MS
28	Martínkovice		+	+	+	+	+	+	+	+	+	+	+			Peichlová 1978, MS
29	Kunštátská kaple 1,2		+	+	+											Rybníčková 1966
30	Velký Děd	3	+	+	+	+	+									Rypl 1980, MS
31	Máj	4	+	+	+											Rypl 1980, MS
32	Úvalno		+		+	+										Opravil 1962
33	Pavlov		+	+	+	+	+				+	+	+	+		Konětopský MS
34	Olbramovice		+	+	+											Svobodová MS
35	Hrabětice										+	+	+	+		Peschke 1977
36	Svatobořice-Mistřín		+	+	+	+	+	+	+	+						Svobodová MS
37	Vracov 1,2	6	+	+	+	+	+	+	+	+	+	+	+			Rybníčková et Rybníček 1972; Sládková-Hynková 1974, MS
38	Malenovice		+	+	+	+										Kneblová-Vodičková 1966b
39	Untersiebenbrunn		+													Havinga 1972
40	Lassee		+	+												Havinga 1972
41	Moosbrunn	1?			+	+	+	+	+	+	+					Havinga 1972
42	Cerová-Lieskové		+	+	+	+	+	+	+	+	+	+		+	+	Krippel 1965
43	Pusté Ulany				+	+	+	+	+	+	+	+	+			Krippel 1963a
44	Dunakeszi				+	+	+	+	+	+	+	+	+	+	+	Járai-Komlódi 1968
45	Ocsa		+	+	+	+	+	+	+	+	+	+				Járai-Komlódi 1968
46	Alpár-Töserdö		+	+	+	+	+	+	+							Járai-Komlódi 1968
47	Ivančiná				+	+	+	+	+	+						Krippel 1974
48	Zlatnická dolina	5	+	+	+	+	+	+	+	+	+					Rybníčková MS
49	Bobrov	13	+	+	+	+	+	+	+	+	+	+	+			Rybníčková MS
50	Jedlová		+	+	+	+	+	+	+	+	+	+				Rybníčková MS
51	Suchá hora		+	+	+	+	+	+	+	+	+					Rybníčková MS
52	Štrbské pleso		+	+	+	+	+	+	+	+	+	+				Krippel 1963b
53	Tatranský domov			+	+	+	+	+	+	+	+	+	+			Krippel 1963b
54	Hozelec	2			+	+	+	+	+	+	+	+	+	+	+	Jankovská MS
55	Spišská Belá-Podhorany				+	+	+	+	+	+	+	+				Jankovská 1972
56	Trojrohé pleso	5	+	+	+	+	+	+								Hüttemann 1983, MS
57	Tokár		+	+	+	+	+	+								Járai-Komlódi et Simon 1971
58	Kökapu				+	+	+									Járai-Komlódi et Simon 1971
59	Rad		+	+	+	+	+									Krippel MS
60	Horeš		+	+	+	+	+	+								Krippel MS
61	Hypkaňa		+	+	+	+	+	+	+	+	+	+				Krippel 1971
62	Vinné												+	+	+	Krippel 1971
63	Smerek 1,2	2	+	+	+	+	+	+	+	+	+					Ralska-Jasiewiczowa 1980
64	Tarnawa Wyżna 1,2	8	+	+	+	+	+	+	+	+	+	+	+	+	+	Ralska-Jasiewiczowa 1980

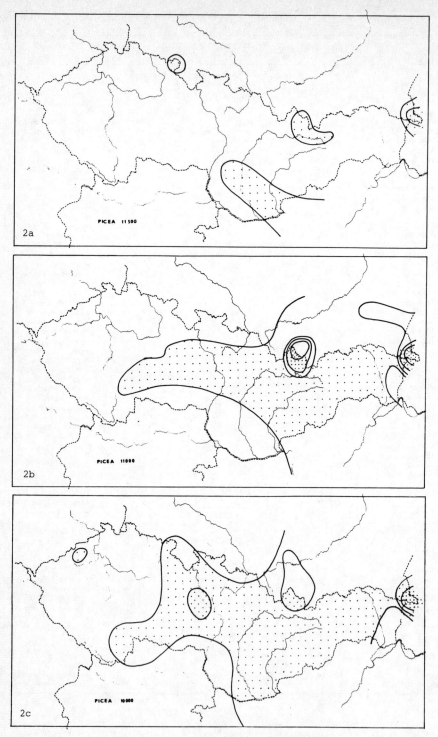

PICEA 11500

2a

PICEA 11000

2b

PICEA 10000

2c

Figure 2 (a-c). Isopollen maps of Picea.

54

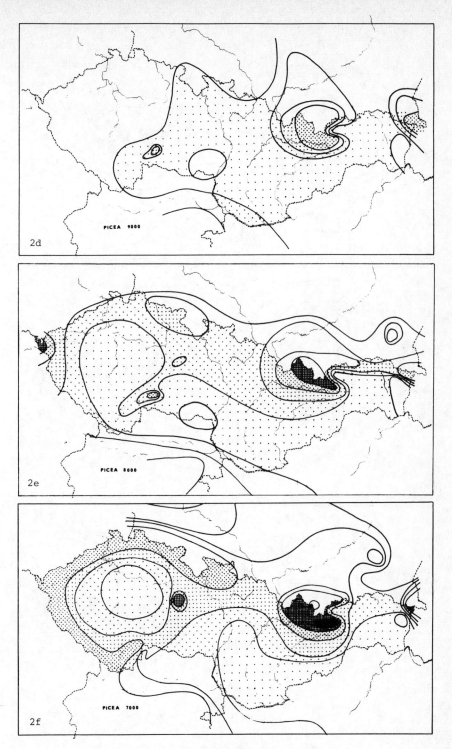

PICEA 9000

2d

PICEA 8000

2e

PICEA 7000

2f

Figure 2 (d-f). Isopollen maps of Picea.

55

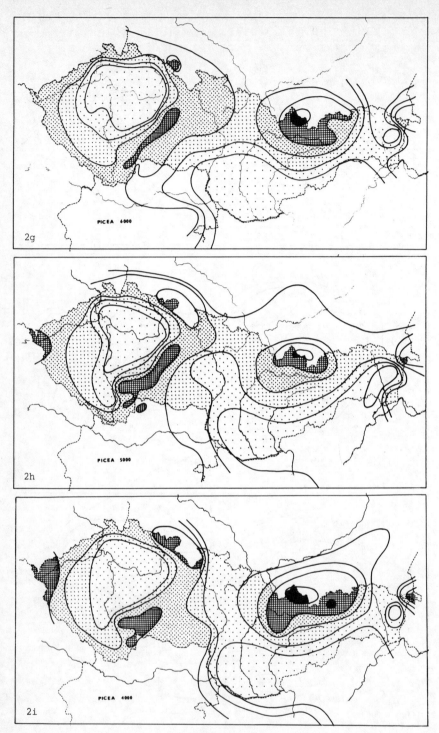

PICEA 6000

2g

PICEA 5000

2h

PICEA 4000

2i

Figure 2 (g-i). Isopollen maps of Picea.

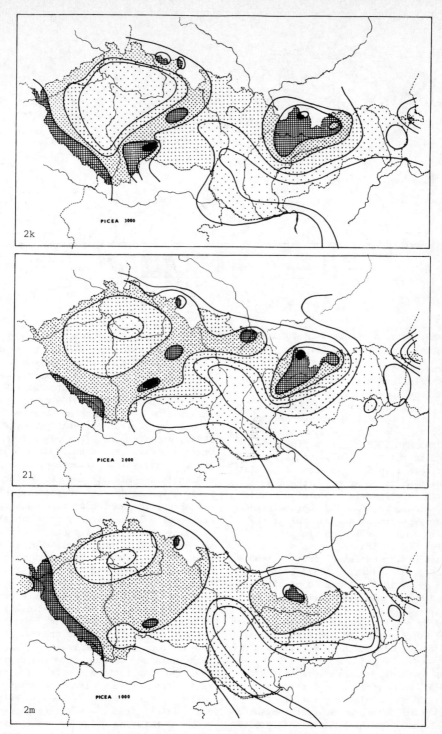

PICEA 3000

2k

PICEA 2000

2l

PICEA 1000

2m

Figure 2 (k-m). Isopollen maps of Picea.

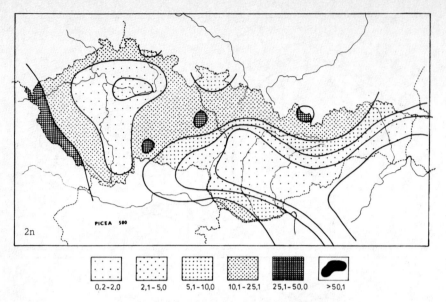

PICEA 500

| 0,2-2,0 | 2,1-5,0 | 5,1-10,0 | 10,1-25,1 | 25,1-50,0 | >50,1 |

2n

Figure 2 (n). Isopollen map of Picea.

500 yrs BP, those of Fagus from 7000 yrs BP, and of Abies from 5000 yrs BP.

3.1 Picea (Fig.2)

Late Glacial and Holocene finds of Picea represent the species Picea abies s.l. in Czechoslovakia. The oldest Quaternary proofs of this species, both macro- and microscopic, come from the Günz and Günz/Mindel (KRIPPEL 1962); most of the finds are dated to the Mindel-Riss and Riss-Würmian interglacials and the Würmian glacial of southern Moravia, of the Ostrava Basin, and of central Slovakia. Special attention must be paid to the finds of spruce needles; they appeared in several calcareous tuffs of Würmian age in the vicinity of Ružomberok and in the foothills of the High Tatra Mts, the latter from Riss-Würmian interglacial (NĚMEJC 1937, 1944; Kneblová 1958, 1960). A survey of all Pleistocene spruce finds was published by OPRAVIL (1978: 101-103). His list has to be enlarged by our own pollen analysis (spruce to 3% of TS) from buried peat in the village of Bulhary in southern Moravia and dated by C14 (Hv 10 855 : 25 875, + 2750, - 2045 BP) to Paudorf interstadial (RYBNÍČKOVÁ 1985), and by spruce pollen finds close to the Czechoslovak border in the Bieszczady Mts (the East Carpathians), dated to about 17 000 BP (RALSKA-JASIEWICZOWA 1980).

OPRAVIL (1978: 110-115) and SAMEK (1973) have also produced a survey of Late Glacial and Holocene occurrence of spruce, published before 1976. The list of their data can be enlarged by recent pollen analyses from Late Glacial of the Bieszczady Mts (1-2% of TS, RALSKA-JASIEWICZOWA 1980), from the Hungrian Plain (8-12% of TS, JÁRAI-KOMLÓDI 1968), from southern Moravia at Vracov (1-2% of TS, SLÁDKOVÁ-HYNKOVÁ 1974), and from the Orava Basin in NW Slovakia (to 2% of TS, RYBNÍČKOVÁ, unpublished). The relatively low percentages of spruce pollen in the Pleistocene, Late Glacial and Early Holocene do not necessarily indicate the absence or ambiguous representation of the species in the areas; they can also be explained by small pollination activity due to severe conditions, similarly, as at present, the pollination decreases with increasing altitude and latitude (cf. SVOBODA,P. 1953).

Isopollen maps show that spruce was distributed over most of Czechoslovakia, except for the driest lowlands, though we do not suppose its monotypic stands. It spread over the country from at least four more or less isolated centres: from the basins below the High Tatra Mts from the Broumov Basin, probably also from the neighbouring basins of the Bieszczady Mts in Poland and , later, from the southern part of the Bohemian-Moravian Uplands. The specific spruce populations or forms (cf. MEZERA 1939; ROUDNÁ 1972; OPRAVIL 1978) could develop in these isolated centres. From out-

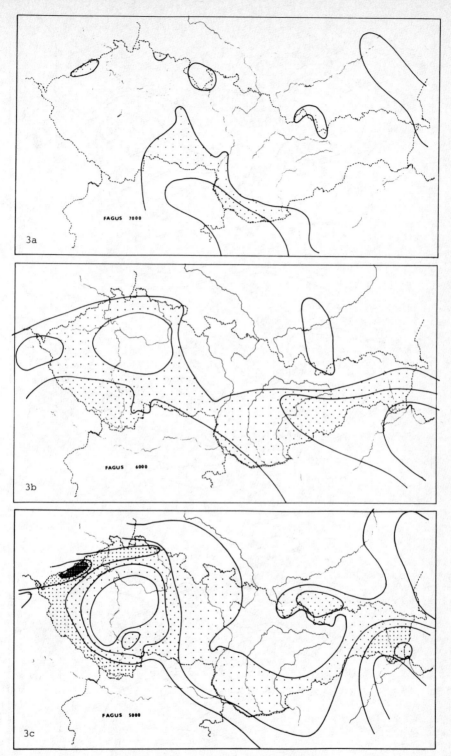

Figure 3 (a-c). Isopollen maps of Fagus.

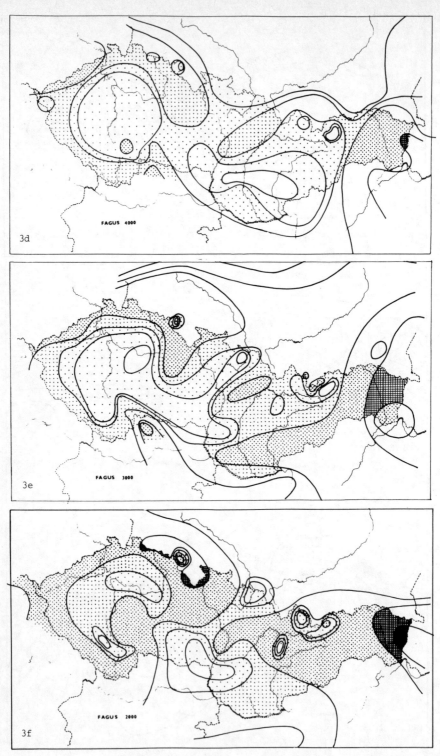

FAGUS 4000

3d

FAGUS 3000

3e

FAGUS 2000

3f

Figure 3 (d-f). Isopollen maps of Fagus.

60

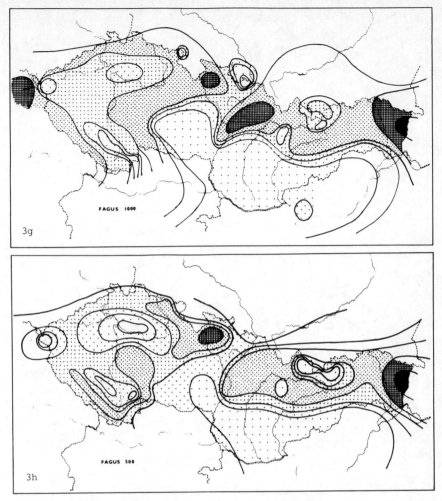

Figure 3 (g+h). Isopollen maps of Fagus.

side the country the isopollen maps show
the main migration routes leading from the
SE, i.e. from the Great Hungarian Plain
and, later on, at about 8000 BP, from the
west or southwest (Fichtelgebirge). The
highest extent of spruce can be found between about 6000 and 5000 (4000) yrs BP,
i.e. during the Atlantic Period, when the
species was the main and probably dominant
tree in the mountain areas of Czechoslovakia (above ca 600 m). After a short re-
treat at about 3000 BP new natural spreading of Picea can be observed from 2000 BP;
this has recently been supported by artificial cultivation and by introduction of
new imported populations (during about
last 150-200 years).

3.2 Fagus (Fig.3)

Pollen of Fagus was produced by Fagus
sylvatica; in the east (in the Carpathians)
we also suspect an admixture of some intermediate populations between Fagus sylvatica and Fagus orientalis (PAGAN 1968).
At present, each is supposed to be a climax
tree of mountain areas over ca 500 m, it
is, however, absent from areas with extreme
temperature regimes and late frosts, such
as in the intermontane basins of the Carpathians.

Although OPRAVIL (1969) brings several
proofs of beech through the Pleistocene
(among them the most interesting find of
charcoal in a palaeolithic settlement at
Dolní Věstonice, southern Moravia,

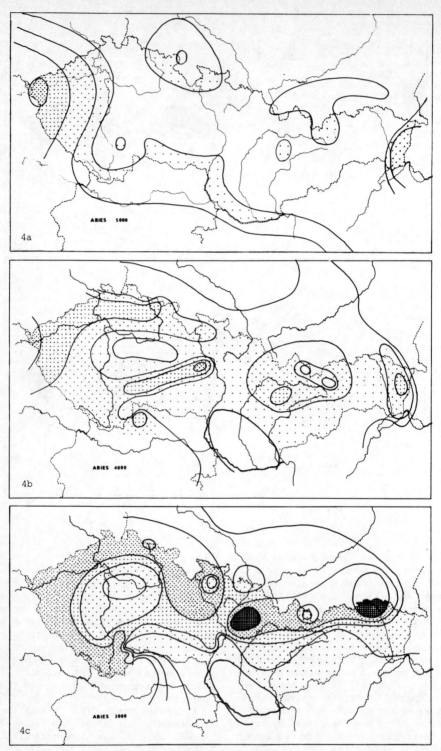

Figure 4 (a-c). Isopollen maps of Abies.

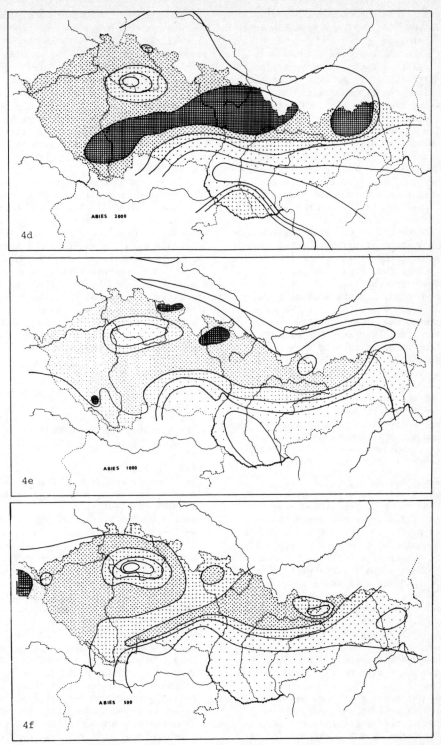

Figure 4 (d-f). Isopollen maps of Abies.

which has been dated to Würmian 3 stadial by KNEBLOVÁ 1954) they are rather scattered on the whole.

During the Holocene the first isolated pollen grains of Fagus in our profiles appeared after 8000 yrs BP. The first iso-pollen lines can be constructed for 7000 BP; at that time the main immigration route led from SE and E(from the Great Hungarian Plain and from the East Carpathians) to the West Carpathians. A scattered occurrence of pollen is observed in the High Tatra Mts and in the Krkonoše Mts; they can, however, come from long distance transport. At 5000 BP the forming of the present distribution patterns of Fagus in the area took place; it came to an end at about 2000 BP, when beech reached its maximum distribution. The area of beech seems to be divided into the two following parts: the Carpathian part, where immigration from the east is supposed, and the Hercynic part, where western populations of beech migrating from the SW seem to be the source. The isopollen maps of Fagus for 2000, 1000 and 500 years BP show partial or total absence of this tree in the Czechoslovak basins, i.e. in the České Budějovice Basin, in the Cheb Basin, in the Broumov Basin and in the intermontane basins of the central West Carpathians. The area of maximum beech pollen concentration well corresponds to a typical beech region in vulcanic mountains of eastern Slovakia (Vihorlat), where - even today - beech is the absolutely dominant tree.

3.3 Abies (Fig.4)

Pollen of Abies in Czechoslovakia has been produced by Abies alba during the Quaternary. The present distribution of the tree covers the area of beech except for the easternmost part of Slovakia (Vihorlat), but it also descends to lower altitudes, forming so called "black coniferous forests" and dividing - in some parts of the country - the beech and oak (oak-hornbeech) zones between (300) 400-500 m a.s.l. (cf. RYBNÍČEK et RYBNÍČKOVÁ 1973).

The isopollen maps of Abies can be constructed since ca 7000 BP. The two oldest maps show the scattered distribution over the area with medium pollen concentration in the western and eastern part of the U.S.S.R. The first published map of Abies for 5000 yrs BP (Fig.4a) shows the main direction of tree migration from SW. Later on, at 3000 and 2000 BP, the forming of the present areas of distribution can be observed. They were similar to those of

Fagus, which at that time was divided into the western and eastern parts along the line of the Moravian Gate and the Morava river valley at 3000 yrs BP. Again, the eastern part was probably formed by other southeastern (Balkan) populations. Later on, at 2000 yrs BP, the two parts of the area seemed to fuse; at this period the maximum occurrence of fir in Czechoslovakia can be found. Since that time Abies alba has retreated, at present, due to increasing air and soil contamination, it is in a calamitous state.

Although we have several finds of Abies alba remains from all interglacials and from the Würmian glacial (OPRAVIL 1976:48-49) and even from the early Holocene (cf. OPRAVIL, op.cit. and SAMEK 1967), we do not believe in the existence of the so-called Proto-Matra refugium in northern Hungary (cf. KRIPPEL 1981). More pollen of fir must be found before 4-5000 BP to substantiate the idea of that refugium. Fir gives an example of explosive spreading.

Isopollen maps of other trees, namely of Quercus, Tilia, Carpinus, Ulmus, Alnus and Corylus are being prepared.

4 REFERENCES

BEUG,H.-J., 1981: Isochrone maps for West Germany. - Maps to the lecture held on the Symposium "The Glacial Refugia and the Migration Paths of Central and North European Trees", Løvenholm 1981, 19 pp. MS.

BIRKS,H.J.B., 1985: A pollen-mapping project in Norden for 0-13 000 BP. - Universitetet i Bergen, Botanisk Institutt, Rapport 38: 1-41.

FIRBAS,F., 1937: Der pollenanalytische Nachweis des Getreidebaus. - Z. Bot. 31: 447-478.

FIRBAS,F., 1949: Waldgeschichte Mitteleuropas. - Jena, 480 pp.

HAVINGA,A.J. 1972: A palynological investigation in the Pannonian climate region of lower Austria. - Rev. Palaeobot. Palynol. 14: 319-352.

HUNTLEY,B. & H.J.B.BIRKS, 1983: An atlas of past and present pollen maps for Europe: 0-13 000 years ago. - Cambridge, 667 pp.

HÜTTEMANN,H., 1983: Postglaziale Vegetationsgeschichte der Hohen Tatra, des Riesengebirges und der östlichen Zentralalpen im Schwankungsbereich der alpinen Waldgrenze. - Dissertation an der Naturwissenschaftlichen Fakultät der Universität Innsbruck, 146 pp. MS.

JANKOVSKÁ,V., 1972: Pyloanalytický příspě-

vek ke složení původních lesů v severo-
západní části Spišské kotliny. - Biоló-
gia 27: 279-292.

JANKOVSKÁ,V. 1980:Palaeogeobotanische Rekonstruktion der Vegetationsentwicklung
im Becken Třeboňská pánev während des
Spätglazials und Holozäns. - Vegetace
ČSSR, A 11, 1-151.

JÁRAI-KOMLÓDI,M. 1968: The Late Glacial
and Holocene flora of the Hungarian
Great Plain. - Ann. Univ. Sci. Budapest.
Sect. Biol. 9-10, 200-225.

JÁRAI-KOMLÓDI,M. & T.SIMON, 1971: Palynolo-
gical studies on swamps of the Zemplen
Mountains. - Ann. Univ. Sci. Budapest.
Sect. Biol. 13, 103-113.

KNEBLOVÁ,V., 1958: Flora interglacjalna w
Ganowcach, wschodnia Slowacja. - Acta
Biol. Cracow, Ser. Bot. 1,1-5.

KNEBLOVÁ,V., 1960: Paleobotanický výzkum
interglaciálních travertinů v Gánovcích.
- Biol. Pr. 6/4, 1-42.

KNEBLOVÁ-VODIČKOVÁ,V., 1966a: Das Spätglazial im Moor bei Zalíbené in Ostböhmen.
- Preslia 38, 154-167.

KNEBLOVÁ-VODIČKOVÁ,V., 1966b: Paleobotanický výzkum rašeliniště v Beskydech. -
Věstn. Ústř. Úst. Geol. Praha 41, 271-278.

KRAL,F., 1979: Spät- und postglaziale Waldgeschichte der Alpen auf Grund der bis-
herigen Pollenanalyse. - Veröff. Inst.
Waldbau Univ. Bodenkult. Wien, 175 pp.

KRIPPEL,E., 1962: Príspevok k problému
floristickej hranice terciér-kvartér. -
Geol. Pr. 63, 157-162.

KRIPPEL,E., 1963a: Postglaziale Entwicklung
der Vegetation des nördlichen Teiles der
Donauebene. - Biológia 18, 730-742.

KRIPPEL,E., 1963b: Postglaciálny vývoj
lesov Tatranského národného parku. - Biol.
Pr. 9/5, 1-41.

KRIPPEL,E., 1965: Postglaciálny vývoj lesov
Záhorskej nížiny. - Biol. Pr. 11/3, 1-99.

KRIPPEL,E., 1971: Postglaciálny vývoj vege-
tácie východného Slovenska. - Geogr. Čas.
23, 225-241.

KRIPPEL,E., 1974: Rekonštrukcia rastlinnej
pokrývky Turčianskej kotliny na základe
pelovej analýzy. - Geogr. Čas. 26, 42-
53.

KRIPPEL,E., 1981: Glaciálne refúgia, post-
glaciálna migrácia a rozšírenie jedle
bielej (Abies alba Mill.) v Západných
Karpatoch. - Geogr. Čas. 33, 133-144.

LOSERT,H., 1940: Beiträge zur spät- und
nacheiszeitlichen Vegetationsgeschichte
Innerböhmens. I. der Kommerner See. III.
Das Spätglazial von Lissa-Hrabanov. -
Beih. Bot. Zbl. 60, 346-394, 415-436.

MEZERA,A., 1939: Rozšíření šiškových forem
smrku v ČSR. - Sborn. Výzk. Úst. Zeměd.
ČSR Praha 50, 1-114.

NĚMEJC,F., 1937: Paleobotanický výzkum
travertinových uloženin Slovenského
krasu. - Rozpr. Čes. Akad. Věd. Um.,Tř.2.
Mat.-Přírod. Praha 46/20, 1-13.

NĚMEJC,F., 1944: Výsledky dosavadních výz-
kumů paleobotanických v kvartéru západního
dílu Karpatského oblouku. - Rozpr. Čes.
Akad. Věd Um., Tř. 2. Mat.-Přírod, Praha
53/35, 1-47.

OPRAVIL,E., 1962: Stáří rašeliniště u Úval-
na, okres Bruntál. - Přírod. Čas. Slez.
23, 225-231.

OPRAVIL,E., 1969: O rozšíření buku (Fagus
silvatica L.) v československém kvartéru.
- Pr. Odb. Přír. Věd Vlastivěd. Úst.
Olomouc 15, 3-59.

OPRAVIL,E., 1976: Jedle bělokorá (Abies
alba Mill.) v československém kvartéru. -
Acta Mus. Siles., Ser. Dendrolog. 25,
45-67.

OPRAVIL,E., 1978: Die Fichte (Picea Dietr.)
im tschechoslowakischen Quartär. - Acta
Mus. Siles., Ser. Dendrolog. 27, 97-123.

PACLTOVÁ,B., 1955: In: PACLTOVÁ,B. & K.MRÁZ:
Kvarterně paleontologické metody studia
dějin lesa a jejich praktické využití. -
Anthropozoikum 5, 365-380.

PACLTOVÁ,B., 1957: Rašeliny na Černé hoře
a dějiny lesa ve východních Krkonoších.
- Ochr. Přír., 12, 65-83.

PACLTOVÁ,B., 1972: Palynologický výzkum k
dějinám rašeliniště ve Františkových
Lázních a jeho osídlení. - Památ. Archeol.
63, 421-428.

PAGAN,J., 1968: Premenlivost morfologických
znakov listov buka na Slovensku. - Zborn.
Vied Prirod.Les. Fak. VŠLD Zvolen 10/1,
15-39.

PEICHLOVÁ,M., 1978: Historie vegetace
Broumovska. - Kand. Disert. Bot. Úst.
ČSAV Průhonice-Brno, 122 pp.

PESCHKE,P., 1977: Zur Vegetations- und Be-
siedlungsgeschichte des Waldviertels
(Niederösterreich). - Mitt. Komm. Quar-
tärforschung Österr. Akad. Wissensch.
Wien 2, 1-84.

RALSKA-JASIEWICZOWA,M., 1980: Late-Glacial
and Holocene vegetation of the Bieszczady
Mts (Polish Eastern Carpathians). -
Warszawa, Kraków, 202 pp.

RALSKA-JASIEWICZOWA,M., 1983: Isopollen
maps for Poland: 0 - 11 000 years B.P. -
New Phytol. 94, 133-175.

ROUDNÁ,M., 1972: Morfologická proměnlivost
původních populací smrku v různých ob-
lastech Československa. - Rozpr. ČSAV,

Ř.Mat. Přír. 82/4, 1-98.

RYBNÍČEK,K. & RYBNÍČKOVÁ,E., 1978: Palynological and historical evidences of vir-
gin coniferous forests at middle altitudes in Czechoslovakia. - Vegetatio 36,
95-103.

RYBNÍČKOVÁ,E., 1964: In: MORAVEC & RYB-
NÍČKOVÁ: Die Carex davalliana-Bestände
im Böhmerwaldvorgebirge, ihre Zusammen-
setzung, Ökologie und Historie. -
Preslia 36, 376-391.

RYBNÍČKOVÁ,E., 1966: Pollen-analytical Reconstruction of Vegetation in the Upper
Regions of the Orlické hory Mountains,
Czechoslovakia. - Folia Geobot. Phytotax.
1, 289-310.

RYBNÍČKOVÁ,E., 1973: Pollenanalytische Un-
terlagen für die Rekonstruktion der ur-
sprünglichen Waldvegetation im mittleren
Teil des Otava-Flussgebietes im Böhmerwaldvorgebirge. - Folia Geobot. Phytotax. 8, 117-142.

RYBNÍČKOVÁ,E., 1974: Die Entwicklung der
Vegetation und Flora im südlichen Teil
der Böhmisch-Mährischen Höhe während des
Spätglazials und Holozäns. - Vegetace
ČSSR, A 7, 1-163.

RYBNÍČKOVÁ,E. & K.RYBNÍČEK, 1972: Erste Ergebnisse paläogeobotanischer Untersuchungen des Moores bei Vracov, Südmähren.
- Folia Geobot. Phytotax.7, 285-308.

RYBNÍČKOVÁ,E, 1985: In: HAVLÍČEK,P. & J.KO-
VANDA: Nové výzkumy kvartéru v okolí Pav-
lovských vrchů. - Antropozoikum 16, 21-
59.

RYBNÍČKOVÁ,E. & K.RYBNÍČEK, 1985: Palaeogeo-
botanical evaluation of the Holocene profile from the Řežabinec fish-pond. - Folia Geobot. Phytotax. 20, 419-437.

RYBNÍČKOVÁ,E., K.RYBNÍČEK & V.JANKOVSKÁ,
1975: Palaeoecological investigations of
buried peat profiles from the Zbudovská
blata marshes, southern Bohemia. - Folia
Geobot. Phytotax. 10, 157-178.

RYPL,R., 1980: Pylové analýzy hřebenových
poloh Hrubého Jeseníku. - Dipl. Pr. Lés.
Fak. Vys. Šk. Zeměd. Brno, 55 pp. MS.

SAMEK,V., 1967: O šíření jedle bílé (Abies
alba Mill.) v době poledové na území
střední Evropy. - Lesn. Čas. 13, 659-666.

SAMEK,V., 1973: O šíření smrku (Picea
abies (L.)Karst.) v době poledové ve
střední Evropě. - Pr. Výzk. Úst. Lesn.
Hosp. Mysl. Zbraslav-Strnady 43, 219-234.

SLÁDKOVÁ-HYNKOVÁ,H., 1974: Paleogeobota-
nická studie rašeliniště u Vracova. -
Dipl. Pr. Kat. Biol. Rostl. Univ. Brno,
74 pp. MS.

SVOBODA,A.M., 1972: Proměnlivost listů buku
lesního (Fagus sylvatica L.) - Studie
ČSAV, Praha 1972/2, 1-143.

SVOBODA.,P., 1953: Lesní dřeviny a jejich
porosty I. - Praha,412 pp.

SZAFER,W., 1935: The significance of isopollen lines for the investigation of
the geographical distribution of trees
in the Post-Glacial period. - Bull. Acad.
Polon. Sc. Lett., Ser. B., 235-239.

Lake, Mire and River Environments, Lang & Schlüchter (eds)
© 1988 Balkema, Rotterdam. ISBN 90 6191 849 9

Pecularities of sedimentation in the small Estonian lakes

L.Saarse
Institute of Geology, Academy of Sciences ESSR, USSR

Abstract: The sedimentation in the Estonian lakes depends on tectonic pattern, climatic conditions, lithology of sediments in the drainage area, geological-geomorphological features of the lake basin and hydrochemical conditions. According to palynological data the accumulation of lake deposits started in South Estonia during Older Dryas, in North Estonia at the end of the Allerød. Late-Glacial minerogenic deposits consist of hydromicas and quartz with kaolinite, chlorites, mixed-layer clay minerals, feldspar, calcite and dolomite, Holocene organogenous and chemogenous deposits are represented by quartz, calcite, dolomite, aragonite, pyrite, siderite, gypsum, feldspar and clay minerals. In Holocene deposits the early diagenetic changes have been proved. The pecularities of sedimentation in regions with different landscapes and some remarks on the water level changes have been given.

The sedimentation in lakes depends on tectonic pattern, climatic conditions, lithology of sediments in the drainage area, geological-geomorphological features of the lake basin and hydrochemical conditions (STRAKHOV 1951). Even in such a small territory as Estonia all above mentioned factors, including tectonic, have influenced the formation of lake deposits. A noticeable role has neotectonic uplift of the earth's crust, due to which numerous coastal lakes have isolated in the northern and western parts of Estonia. Among them the oldest geologically investigated lake is Kahala, that became an isolated lake at the end of the Preboreal period and the youngest one is Lake Käsmu, which was separated in the middle of V stage of the Limnea Sea, approximately 800 years B.P.

Sedimentation in lakes was highly affected by the climatic conditions, which can be seen in the alternation of deposits of various composition. According to palynological data (PIRRUS 1969; PIRRUS & RÖUK 1979) the accumulation of lake deposits started in South-Estonia during Older Dryas, in North-Estonia at the end of Allerød. In Older Dryas the brownish-grey and grey minerogenic deposits were formed. In comparison with Older Dryas limnoglacial var-

ved clays reaching the thickness of 18 m (usually 3-8 m), the thickness of minerogenic lake deposits of the similar age extends to 2 m. The overlaying Allerød lake deposits are represented by grey silt and clay often with black-grey interbeds of low content of organic matter with thickness of about 1-3 m. The upper part of late-glacial lake deposits, the representative of which is silt, often with fine-grain sand interlayers, containing remnants of Bryales moss, was formed in Younger Dryas. The thickness of these deposits ranges from 0.5 to 4.0 m. The clayey late-glacial deposits are common only in lakes of hummocky morainic relief of South-Estonia, drumlin field of Saadjärve and morainic plateau of North-Estonia. They are very thin or totally absent in lakes of kame fields and fluvioglacial plains (Illuka, Viitna, Uku etc., Fig.1,2).

The grain-size composition of late-glacial lake deposits is rather uniform, showing that mainly the minerogenic material with silty size was carried into the lakes. In drainage lakes and in the littoral zone minerogenic material is coarse. Out of the three types of granulometric spectra the most frequent are deposits with $\emptyset$ 0.05-0.01 mm, which were formed during Older Dryas and also Younger Dryas. The second type of gra-

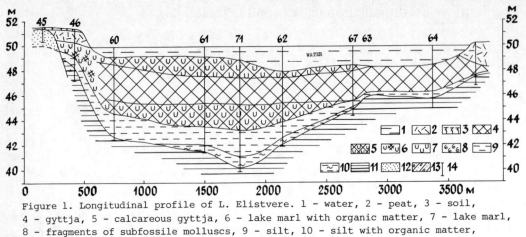

Figure 1. Longitudinal profile of L. Elistvere. 1 - water, 2 - peat, 3 - soil,
4 - gyttja, 5 - calcareous gyttja, 6 - lake marl with organic matter, 7 - lake marl,
8 - fragments of subfossile molluscs, 9 - silt, 10 - silt with organic matter,
11 - varved clay, 12 - sand, 13 - till, 14 - borings.

nulometric spectra with high amount of fine silt and clay fractions is characteristic of Allerød beds. Differently from the grain-size of the Estonian varved clays, the composition of which depends on bedrock composition, on the size, morphometry and topography of sedimentational basins, the grain-size composition of lake deposits (Table 1) is rather similar to the different bedrock outcrops and landscape regions. It shows that fine material was carried into small isolated lakes, grain-size of which depended on the kinetic energy of the transportational force.

According to the X-ray diffractograms the late-glacial minerogenic deposits of Estonia consist of hydromicas and quartz, to smaller extent of kaolinite, chlorites, mixed-layer clay minerals, feldspar, calcite and dolomite.

The content of calcite and dolomite generally extends up to 5 - 10%, whereas it in the underlaying limnological deposits extends up to 20-30%. At the same time the role of dolomite as well as the ratio MgO/

CaO is increasing towards the south. A different degree of crystallization of chlorites from the deposits of comparatively warm and cool intervals of Late-Glacial period has been found (PIRRUS & SAARSE 1978). In the lake deposits of Saadjärve drumlin field chlorites, established during Allerød period, have a higher degradation degree. In comparison with the degree of degradation of lacustrine and limnoglacial deposits, it should be stressed that minerals in the first one are more degraded, showing that these in unconsolidated, saturated with water late-glacial lake deposits have had no opportunity to form more suitable crystalline forms.

Holocene lake deposits are represented by gyttja and lake marl, which started to accumulate in Preboreal period (Fig.3). Calcareous lake deposits are rather widespread in lake sequences (Fig.5). The preconditions for calcite accumulation were climatic conditions, hydrochemical peculiarities in the sedimentational basins, lithology of sediments in the catchment

Table 1. Grain size composition of Late-Glacial and Holocene lake deposits of Estonia, %.

Type of deposits	>0.1	0.1-0.05	0.05-0.01	0.01-0.005	0.005-0.002	0.002-0.001	<0.001
Late-glacial limno-glacial deposits	3.4	3.8	56.6	15.5	4.7	8.8	9.2
Late-glacial lake deposits	9.0	9.3	51.3	10.5	8.0	4.8	7.1
Holocene lake deposits	11.0	12.2	43.9	7.0	7.6	6.9	11.4

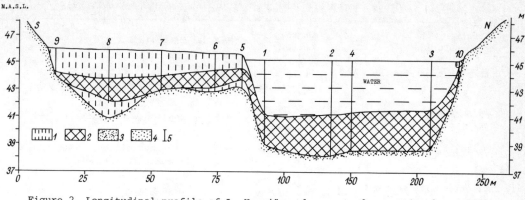

Figure 2. Longitudinal profile of L. Haugjärv. 1 - peat, 2 - gyttja, 3 - gravel,
4 - sand, 5 - borings.

area and carbonate bedrock near the soil
surface. The most intensive accumulation
of calcareous lake deposits took place
in Boreal and Atlantic periods, but it
varies greatly in different lakes (MÄNNIL
1967). According to X-ray data they mainly
consist of calcite with low admixture of
dolomite, hydromica, kaolinite and pyrite
(Fig.4).

Organogenous deposits (the Holocene lake
deposits with organic matter more than
15%) contain various amounts of ash (15-85%
average being 30-40%). Fine detritus gyttja
with ash content less than 35% has accumulated in enclosed lakes or in lakes with
small influx, sandy deposits in the catchment area (Fig.6). Gyttja with medium ash
content (35-50%) accumulated in drainage
lakes and lakes with till beds in their
catchment area. Enriched ash containing
gyttja (over 50%) is characteristic of the
drainage lakes in river valleys (Viljandi,
Kariste) and lower part of lake sequences,
formed in the condition of intensive erosion at the beginning of Holocene. The
tendency of distribution of ash content in
profiles is various, ordinary with lower
amount in the middle part of sequences,
which were formed during the warm Atlantic
period. The other chemical compounds are
also different: SiO_2 content ranges from
1.30 to 66.11%, Al_2O_3 - 1.30-10.06%, Fe_2O_3
- 0-34% (Fig.7), CaO - 0-31%, MgO - 0.1-
5.5%. The highest amount of K_2O and Na_2O
(2.56 and 0.67%) is in sediments enriched
with terrigenous substances in the lower
part of sequences due to intensive erosio-

nal processes and in the upper part of sequences, induced by human impact (slash-
and-burn agriculture), which also favoured
the erosion (Fig.6). The content of P_2O_5
in lake deposits is usually smaller than 1%,
but sometimes it extends up to 2.5%. In
these cases the brushite $CaH(PO_4) \cdot 2H_2O$ on
diffractograms has been identified.

Using X-ray techniques the mineralogical
composition of gyttja has been established.
For the identification of minerals from un-
oriented samples the heating treatment, HCl
and HF was used (Fig.8). Taking advantage
of computer techniques the chemical composition of gyttja was also calculated. The
following minerals have been ascertained
from the bottom deposits of the Estonian
lakes: quartz, calcite, dolomite, aragonite, pyrite, siderite, gypsum, feldspar,
hydromicas, mixed layer minerals, chlorites
and kaolinite (SAARSE et al. 1984).

On the basis of the Hewlett-Packard CHN-
analyser the elemental composition of organic matter was studied. In all of the
samples carbon (C) prevails (24 - 66%) over
the content of hydrogen (H 1.5-6.3%) and
nitrogen (N 0.6-3.5%). It is supposed that
ratio H/C and O/C shows the origin of organic matter - the high quantity of H/C
ratio would indicate the algae derived organic matter, while low H/C and high O/C
ratio would show the presence of terre-
stial plant remnants.

The group composition from organic part
of deposits was determined according to
the following scheme (Table 2).

69

Table 2. Group composition analysis according to PALU & VESKI, 1982

Extraction with methanol benzene solution	-->	Organic matter	-->	Bitumoid A
Hot water treatment	-->	Residue I	-->	Hot water extract (sugars, strach, etc.)
Boiling water treatment	-->	Residue II	-->	Boiling water extract
Hydrolysis with 2% HCl	-->	Residue III	-->	Easy hydrolyzable matter (hemicellulose, etc.)
Extraction with methanol benzene solution	-->	Residue IV	-->	Bitumoid C
Treatment with 1% NaOH solution	-->	Residue V	-->	Alkali-soluble matter (necromic matter)
				Humic acids Fulvic acids
Hydrolysis with 80% H2SO4	-->	Residue VI	-->	Difficult-to-hydrolyse matter
		Residue VII	-->	Non-hydrolyzable matter (lignin, etc.)

Table 3. Group composition of the Estonian deposits, %.

Lake		Ash	Water soluble com.	Hemi-cellu-lose	Bitu-moid	Humic acids	Cellu-lose	Non-hydro-lyzable residue
Kirikumäe		4.7	2.8	0.2	8.2	19.6	2.3	22.2
Koobassaare		15.6	2.7	0.2	8.4	29.3	1.5	50.3
Kõverjärve		20.2	1.2	0.1	8.3	45.6	1.6	19.0
Ruila		30.4	2.9	0.2	8.3	20.2	2.6	21.3
Kahala		41.7	5.3	13.4	9.5	13.0	0.5	19.6
Ülemiste		48.9	3.2	9.1	9.5	16.1	1.1	16.0
Lahepera	1.5m	51.3	4.9	30.6	10.0	32.8	2.1	8.9
"	3.0m	68.9	4.8	26.5	8.9	28.8	3.4	11.8
"	3.5m	68.1	4.8	25.1	8.5	29.9	2.9	10.8
"	4.5m	77.0	5.5	22.6	7.6	40.4	1.6	12.6
"	6.5m	55.1	4.1	19.2	10.1	41.0	3.8	10.8
"	8.0m	59.3	3.6	13.4	10.4	42.3	5.8	10.8

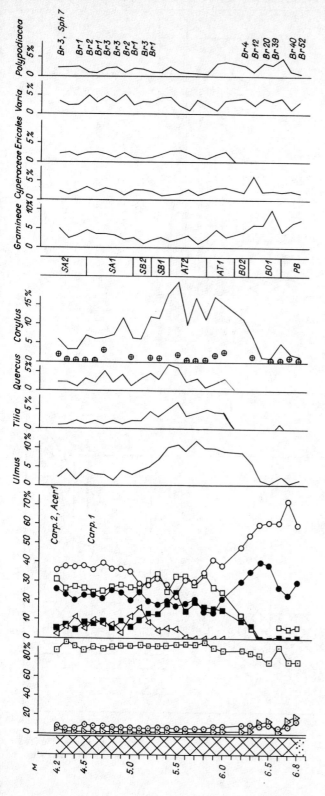

Figure 3. Pollen diagram of L. Haugjärv. Analyzed by A. Sarv.

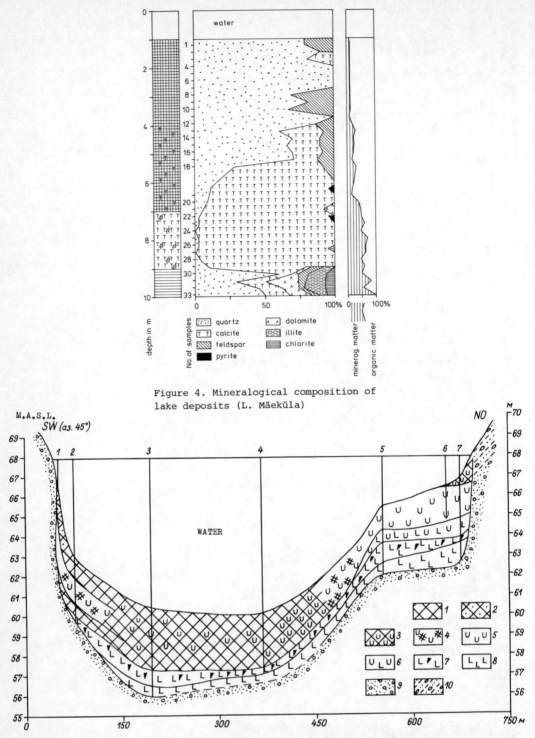

Figure 4. Mineralogical composition of lake deposits (L. Mäeküla)

depth in m · No. of samples

	quartz		dolomite
	calcite		illite
	feldspar		chlorite
	pyrite		

minerog. matter · organic matter

Figure 5. Longitudinal profile of L. Kiruvere. 1 - gyttja, 2 - sandy gyttja, 3 - calcareous gyttja, 4 - gyttja and lake marl interbedding, 5 - lake marl, 6 - silty lake marl, 7 - silt with organic matter, 8 - silt.

72

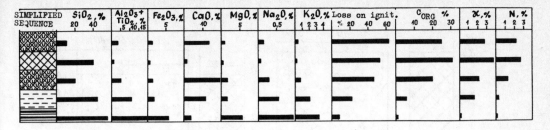

Figure 6. Average chemical composition of the lakes of Saadjärve drumline field.

Some results of group composition of the Estonian lake deposits are presented in Table 3.

Bitumoid was isolated from the dry sediment in Soxhlet extraction using benzene-methanol solution (3:1). Total bitumoid content in Holocene lake deposits ranges from 3 up to 15% with bitumoid A dominating, which surpasses 1.5-2.7 times the content of bitumoid C. It is interesting to point out that in Eemian interglacial sediments the content of bitumoid has decreased in percentage 8-10%, in Holsteinian beds 5-8% and in our Ordovician oil-shale 0.5-1.0%. The amount of humic acids (alkali-soluble compounds) is also very informative: in Lahepera sequence humic acids increase in depth (Table 3), PALU & VESKI 1982). The variability of humic acids has been explained in different ways - by the influence of peat bogs in the catchment area, with the synthesis of carbohydrates etc. We think, there is a possibility that during the diagenesis the easily hydrolyzable compounds turned into humic acids, because correlation with them is rather high (PALU & VESKI 1982). The percentage of non-hydrolyzable residue in Holocene lake deposits is rather low - 9-32% (in one case 50.3%, Table 3), whereas in fuels and oil-shale it always forms over 50%.

The results obtained from investigations on Holocene lake deposits give proof that the following early diagenetic changes in gyttja took place:
- increase in the ratio C/N;
- decrease in the total amount of easily hydrolysed compounds;
- increasing quantity of alkali-soluble (humic) compounds.

The studies of the composition and formation of lake deposits in different landscape regions have also been carried out. In the lakes of fore-glint coastal plain

the thin detritus gyttja, silty and sandy gyttja with low amount of calcareous component have been formed. In the lake basins on the North-Estonian carbonate plateau, filled with calcareous rich gyttja or lake marl, sedimentation started in Preboreal or at least in Boreal. Dystrophic lakes with a wide distribution of lake marl are common in West Estonia. Kettle lakes in the kame fields are rich in gyttja (Fig.2) with ferruginous compounds in the lower part of sequences (Fig.7) and in rare cases some calcareous-rich interbeds are formed in their littoral zone. Lakes on the Central-Estonian watershed, especially in Saadjärve drumlin field are of specific importance for the solving of stratigraphical problems with their uneven profiles, including the Late-Glacial part of sequences, starting in Older Dryas. Lake sediments on the Middle-Devonian plateau with highly disjointed morainic relief, intersected by deep ancient valleys, are represented by minerogenic gyttja in drainage lakes and organic-rich gyttja in kettle lakes with calcareous facies in the littoral zones. On the Upper-Devonian plateau with morainic hills and hummocks, fluvioglacial and limnoglacial kames on the Haanja Height, with fluvioglacial, alluvial deposits and peat in Võru-Hargla basin and limnoglacial sand in Palumaa area, the organogenous deposits in recent lakes are prevailing.

Taking into account the distribution and frequency of macrophyte pollen (PIRRUS 1969, RÕUK 1979, KESSEL et al. 1982, SAARSE et al. 1985, etc.), fragments of terraces, changes of the bed limits and peculiarities of mollusc fauna (MÄNNIL 1967), the curve, showing the main tendency of water level changes, has been drawn. According to the latter the lowering of the lake water level took place in Early Preboreal, Boreal and Atlantic and in the middle of Subboreal and

73

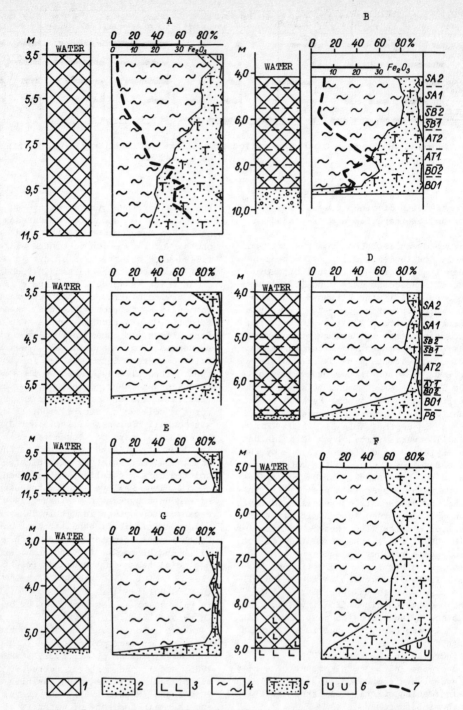

Figure 7. Composition of lake deposits. A – Räätsma, B – Rääkjärv,
C – Ahnejärv, D – Haugjärv, E – Valgejärv, F – Konnjärv, G – Kurtna.
1 – gyttja, 2 – sand, 3 – silt, 4 – organic matter, 5 – terrigenous
matter, 6 – carbonate compounds, 7 – Fe_2O_3 content.

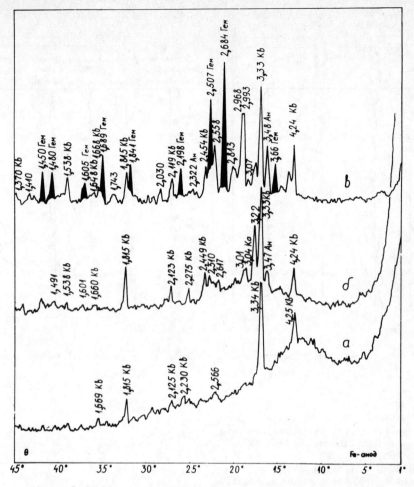

Figure 8. Diffractogram of the lake deposits of L. Mäküla, sample no 6, depth 2.4 m. a - natural, b - after heating at 500°C, c - after heating at 1000°C.

Subatlantic periods, the rising - later part of Preboreal, Late-Boreal and Atlantic periods. In conclusion it would be interesting to point out the fact that the most significant water level changes, which took place in Early and Middle Holocene, coincide with eustatic changes in the Baltic Sea obtained on the territory of Estonia (KESSEL & PUNNING 1984).

REFERENCES

KESSEL.H. & J.-M.PUNNING, 1984: The development of the Baltic Sea in the Holocene. - In: Estonia. Nature, Man, Economy. Tallinn, 36-47.

PIRRUS,R. & A.-M.RÕUK, 1979: Uusi andmeid Soitjärve nõo geoloogiast. - In: Eesti NSV saarkõrgustike ja järvenõgude kujunemine. Tallinn, lk, 118-144.

PIRRUS,R. & L.SAARSE, 1978: Late-Glacial lake deposits in Estonia. - Polskie Archiwum Hydrobiologii, 25, 1/2, 333-336.

KESSEL,H., L.SAARSE, R.SINISALU & K.UTSAL, 1982: Geologicheskoe razvitie ozera Kahala. - Izv. AN ESSR. Geologya, 31, 1, 21-28.

MÄNNIL,R., 1967: Nekotorye cherty osadkonakoplenya v pozdne- i poslelednikovyh ozerah Estonii. - In: Istoria ozer Severo-Zapada. L., 300-305.

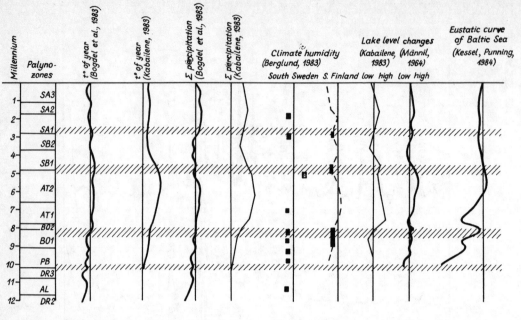

Figure 9. Some paleoclimatic parameters. 1 - warming, 2 - periods of the extreme changes by L. Starkel.

PALU.,V. & R.VESKI, 1982: Ob izmenenii grup-
povovo sostava sapropelya ozera Lahepera
v zavisimosti ot glubiny. - Izv. AN
ESSR. Khimya, 31, 2, 136-140.

PIRRUS,R.: 1969: Stratigraphicheskoe
raschlenenie pozdnelednikovyh otloženii
južnoi Estonii po dannym sporovo-pylt-
sevovo analyza. - Izv. AN ESSR. Khim.
Geol. 18, 2, 181-190.

SAARSE,L., A.SARV & J.KARIN, 1985: Osad-
koobrazovanie i geologicheskoe razvitie
ozer Illukaskovo kamovovo polya. - Izv.
AN ESSR. Geologya 34, 2, 62-68.

SAARSE,L., K.UTSAL & E.LÕOKENE, 1984: Pri-
menenie rentgendifraktometricheskovo me-
toda pri izuchenii veschestvennovo sostava sapropelei. - Izv. AN ESSR. Geologya 33, 1, 19-26.

STRAKHOV,M.N., 1951: Izvestkovo-dolomit-
ovye facii sovremennyh i drevnyh vodoio-
mov. - Tr. IG nauk. M. 124, 45, 372 pp.

Lake, Mire and River Environments, Lang & Schlüchter (eds)
© 1988 Balkema, Rotterdam. ISBN 90 6191 849 9

Principles of the palaeoecological subdivision of the European part of the USSR

L.Saarse & A.Raukas
Institute of Geology, Academy of Sciences ESSR, USSR

D.Kvasov
Institute of Lake Research, Academy of Sciences, USSR

Abstract: The European part of the USSR occupies a large territory with different geological, geomorphological and climatic conditions as well as vegetation. Phytogeographers have distinguished here tundra, woodland tundra, northern, central and southern Boreal (taiga), Boreo-nemoral, Nemoral, woodland Steppe, Steppe and Steppe Desert zones and subzones. In addition to the biotic zonation the natural-geographic subdivision of this area is presented with reference sites studied.

1 INTRODUCTION

The European part of the USSR includes a large territory with the approximate area of 5.2 million sq. kilometres, that surpasses the total area of West-European countries. Because its geological, geomorphological, pedological and climatic conditions, as well as vegetation are very variable, the paleoecological subdivision of such a vast territory is rather complicated.

According to the phytogeographers E.LAVRENKO and V.SOCHAVA (1954) the tundra, woodland tundra, northern, central and southern Boreal (taiga), Boreo-nemoral, Nemoral, woodland Steppe, Steppe and Steppe Desert zones and subzones have been distinguished (Fig.1), which correspond to the biotic zones and subzones by H.SJÖRS (1963).

In addition to the biotic zonation the natural-geographical subdivision of the westernmost part of the territory of the Soviet Union has been worked out taking into account the differences in geological and geomorphological structure and the peculiarities of distribution of soil and plant cover. These single natural-geographical regions meet the demands of type regions, suggested by B.BERGLUND (1979).

2 DESCRIPTIONS

2.1 On the Kola peninsula there are approximately 107 000 lakes in tundra, woodland tundra and northern taiga subzones, characterized by extremly poor mineralization of lakewater (total sum of ions 40-60 mg/l). Most of the lakes are small with their area less than 1 sq.kilometre, they are in oligotrophic stage of development and show distinct features of dystrophic stage, avoiding thus the mesotrophic stage (SEMENOVICH 1958). The variety of lacustrine deposits is rather high. On the Rybachevo peninsula (tundra zone) the lakes are rich in siliceous mud (diatomite), whereas in the coastal area of the Barents Sea terrigenous deposits are dominating. In the central and southern parts of the Kola peninsula there occurs siliceous rich mud with ferromanganese crusts in the profundal part of the lakes formed due to the diatom rich flora and ground water with abundant ferromanganese compounds. According to N.M.STRAKHOV organic matter plays the leading role in the formation of ferromanganese crust and nodules. As the lakes on the Kola peninsula are poor in organic matter, Fe_2O_3 is prevailing and the content of FeO is low. The southern part of the Kola peninsula belongs to the northern taiga subzone, it is very swampy and so the peaty gyttja and dy are common representatives of lake deposits.

2.2 The Karelian ASSR with its 43 000 lakes, area of which comprises a 18% of the territory surpasses in this respect all the other republics of the USSR. Mesotrophic lakes are prevailing with the total sum of ions being 100-130 mg/l in water.

Figure 1. Subdivision of the European part of the Soviet Union.

1: Boundaries between biotic zones; 2: Boundaries between natural geographic regions; 3: Frontier between republics; 4: Reference sites.

I-X: Biotic zones and subzones. I: Tundra; II: Woodland tundra; III: Northern Boreal (northern taiga); IV: Central Boreal (middle taiga); V: Southern Boreal (southern taiga); VI: Boreo nemoral; VII: Northern Nemoral; VIII: Woodland Steppe; IX: Steppe, X: Steppe Desert (semidesert).

Natural geographic regions (type regions):
Ko: Kola peninsula. a: Northern coastal; b: Western region of Murmansk; c: North-western mountains; d: Eastern region of Murmansk; e: Central region; f: Hibins and Lovozero; g: Notozero depression; h: Central mountain; i: Southern depression; j: Kandalaksha; k: Kovdozero Lowland; l: Terski.
Ka: Karelian ASSR. a: North-western mountain; b: Lake district; c: White Sea Lowland; d: West-Karelian Upland; e: Onega-White Sea watershed; f: Äänisjoki-Suojoki; g: Suna-Shuya watershed; h: Polovenets Inlet-River Vyg watershed; i: Ladoga basin; j: Onega basin; k: Onega-Ladoga watershed; l: Vodlozero basin,
E: Estonian SSR. a: North Estonia; b: West Estonia; c: Intermediate Estonia; d: Central Estonian watershed; e: Peipsi-Vortsjärv Lowlands; f: Middle Devonian Plateau; g: Upper-Devonian Plateau.
La: Latvian SSR. a: Coastal Lowland; b: Kurzeme Upland; c: Middle-Gauja Lowland; d: Central-Vidzeme and Aluksne Upland; e: East Latvian Plain; f: Lubano and Middle-Daugava Lowland; g: Latgale Upland.
Li: Lithuanian SSR. a: Baltic coast Lowland; b: Zhemaitiya Upland; c: Middle-Lithuanian Plain; d: West-Auksthaitaya Plateau; e: Baltic ridge; f: South-eastern fluvioglacial Plain.
B: Byelorussian SSR. a: Byelorussian Valdai; b: East Byelorussia; c: East Baltic; d: West Byelorussia; e: Predpolesje; f: Polesje.
U: Ukrainian SSR. a: Volynia; b: Zhitomir; c: Kiev; d: Levoberezhnaya Polesje; e: Volyno-Podolya; f: River Dnepr; g: Levoberezhnaya; h: Central Russian woodland Steppe; i: Northern Pravoberezhnaya; j: Northern Levoberezhnaya; k: Donetskaya; l: Starobelskaya Steppe; m: Black-Sea and Tauria; n: The Carpathians; o: The Crimea.
Mo: Moldavian SSR. a: Woodland Steppe; b: Volyno-Podolya Upland; c: Kodr Upland; d: Kodr; e: Bessarabian Plain; f: Tiche Upland.
R: Russian SFSR, R-a north-eastern lake district: $R-a_1$: Tundra; $R-a_2$: Woodland tundra. R-b north-western lake district: $R-b_1$: Lakes of the Valdaian glaciation in the northern taiga; $R-b_2$:Middle taiga; $R-b_3$: Southern taiga; $R-b_4$: Lakes in the catchment area of the Gulf of Finland. R-c taiga lakes outside of the Valdaian glaciation: $R-c_1$: Northern taiga; $R-c_2$: Middle taiga; $R-c_3$: Southern taiga. $R-d_1$: Lakes in the catchment of Rivers Velikaya-Lovat; $R-d_2$: Lakes on the Valdaian ice marginal zone. R-e: Lakes on the Middle Pleistocene and periglacial area. R-f: Lakes of northern Nemoral zone. R-g: Lakes of woodland Steppe. R-h: Lakes of Steppe. R-i: Lakes of Steppe Desert zones.

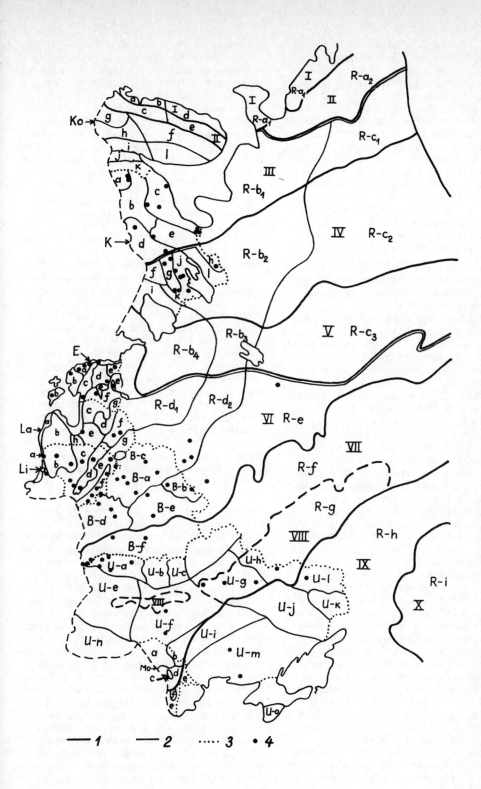

The share of organic matter and organogenous deposits has increased in comparison
with that in the lakes of the Kola peninsula. The main types of lacustrine deposits
are gyttja, peaty gyttja, silty mud with
ferromanganese crusts and nodules in the
upper part of sequences. Terrigenous deposits can be found in the littoral zone of
large lakes. Although several Karelian lakes have been studied, only the overgrown
lakes and mires (altogether 20) have been
suggested as reference sites. According to
available data (ELINA 1981) they include
also the sites, where the sedimentation
started in Allerød (Shombasuo, Gotlavonok,
Dlinnoe), but the outset of the accumulation of the organogenous deposits commonly
falls in Boreal, seldom in Preboreal.

2.3 The lakes of the Baltic Republics and
Byelorussia are most extensively studied.
Their trophic stage varies from oligotrophic to ultraeutrophic and the mineralization degree of lake-water shows differences
as well. The lacustrine deposits are represented by gyttja, peaty gyttja, calcareous
gyttja, lake marl, organogenous lake marl,
silty lake marl, silt, sand and gravel.
The ferrogenous and siliceous-rich deposits
are not typical to this area. At the same
time, this is the region where the precipitation of calcareous mud is taking place
due to the carbonate bedrock and ground
water rich in carbonate compounds. During
the whole Holocene, especially in Boreal
and Atlantic periods the formation of lake
marl took place in the lakes over the large
area farther off the Baltic region - from
the Arctic circle in north up to 50 degree
of northern latitude in south (BARTOSH
1976).

The thickness of Late-Glacial and Holocene lake deposits is remarkable - up to
38.5 m, 6 - 8 m in average. Throughout the
abovementioned Baltic republics Older Dryas
deposits are terrigenous and even in Bølling deposits organic matter is embedded
in negligible amount. Allerød (time span
11 800 - 10 800 B.P.) deposits in Byelorussia and Lithuania serve as indicating
beds, represented by peat or plant remnant
rich terrigenous beds. In North-Baltic region the Allerød beds are terrigenous
consisting of enlarged amount of diffusal
organic matter as here the influence of
retreating continental ice was greater and
disappeared only in Younger Dryas. In Estonia 7 type regions with 22 reference sites,
in Latvia 8 type regions with 3 reference
sites, in Lithuania 6 type regions with 17

reference sites and in Byelorussia 6 type
regions with 20 reference sites have been
selected by L.SAARSE, A.RAUKAS, G.EBER-
HARDS, M.KABAILENE and O.JAKUSHKO correspondingly. In these type sequences the lacustrine sedimentation started in different
times but not earlier than in Older Dryas.
The beginning of the formation of organogenous deposits is also asynchronous, in
southern area it extends to Allerød (peat
interbeds), in northern area to Preboreal.

2.4 In the Ukraine, lakes are located in
the northern part of the republic, in the
territory of Polesje 5 lakes have been suggested as reference sites. Both the formation of those lakes in Polesje and the
accumulation of deposits in them, started
in different times, in lake Svyatoe in
Allerød, in lake Makovich in Younger Dryas.
In the other reference sites (lakes Ostra-
venskoe, Meźlenskoe and Dvirskoe) the formation of gyttja begins only in Boreal.
There are a few lakes in woodland Steppe
and Steppe zones due to rather arid climate
and smooth surface relief, they are located
in river valleys and serve as typical oxbow
basins. So it is quite natural that for
these zones only the mires (10 of them)
have suggested by S.TURLO and M.VEKLICH as
reference sites. The thickness of peat deposits in these mires is 2.8 - 12 m. In the
mires of Zgar, Girlovoe and Troitskye the
accumulation of deposits started in Allerød,
in Vily and Makhavatoe in Atlantic period.

2.5 Taking into consideration geological
and geomorphological peculiarities of the
relief formed during the different glaciations in the territory of the Russian SFSR
the following 8 lake districts (lacustrine
regions) can be differentiated. Lake district of far north-east (R-a) bounded by
the Rivers Usa, Pechora and Tsilma in the
south including also the watershed of the
rivers on the Kanin peninsula and Cheskaya
bay. This area was entirely coated by the
Weichselian ice cover. Ice readvanced from
this area finally in the ninth millenium
B.P. There are numerous lakes in Maloze-
melskaya and Bolshezemelskaya tundra regions
due to permafrost pattern and disjointed
relief, especially in the ice marginal zones. In the coastal area of the Barents
Sea the relic coastal lakes are common.
Their deposits are represented by organogenous gyttja, terrigenous silt, sand and
gravel with ferromanganese varieties in
some lakes.

2.6 Lake district of north-west (R- b) intersects the ice marginal zone of Weichselian glaciation in the line which runs through Pskov, Staraya-Russa and Vyshegorsk and coincides with it in the south-east. In this district several smaller areas should be differentiated, first of all the largest lakes like Onega and Ladoga, the large lakes like Peipsi and Ilmen on the watershed of the Gulf of Finland and Beloe, Kubenskoe, Vože and Lacha on the watersheds of Rivers Sheksna, Sukhona and Onega. The accumulation peculiarities, distribution and composition of bottom deposits and geological history of these lakes have been eluciated, however, they can't be used as reference sites. Due to their large area, they do not meet the demands of the IGCP Project No 158 "Paleohydrology in the temperate zone". Lakes on the lowland in the watershed of the Gulf of Finland (R- b_4) are scattered over the territory west and south of L.Ladoga. On the Karelian isthmus there are approximately 700 lakes. The stratigraphy and the character of bottom deposits have been elucidated for the lakes Krasnoe, Lopata, Glukhoe, Vuoksa, Blagodatnoe etc. (MALYASOVA, SPIRIDONOVA 1965; KVASOV, MALYASOVA 1969) showing that the organogenous sedimentation started in various time span, mostly in Boreal. The depression on L. Krasnoe in canyon, like an ancient valley cuts into crystalline basement down to the depth of 60 m below the sea level and is filled with fluvioglacial, limnoglacial and lacustrine deposits testifying that this basin was blocked by stagnant ice.

Lakes in the ice marginal zone of the Weichselian glaciation (R- b_{1-3}) are mostly assembled on the ice marginal Heights Vepsa, Valdai etc. Most of them are comparatively small with the area less than 0.5 sq. kilometers with medium mineralization of lake water, that greatly depends on the geological-geomorphological structure of an area. Silt, clay, silty and clayey gyttja with ferromanganesian crusts and nodules and clastogene facies in the littoral zone are widespred in these lakes. Southward the role of organogenous deposits is gradually increasing, but this general trend is in places interrupted by the differences in depositional conditions in local lakes. The stratigraphy and geological history of these lakes are cleared up rather deficiently.

2.7 There are a few lakes in the district of the taiga zone, beyond the limits of Weichselian glaciation (R- c). The formation of lake basins has been predetermined by the processes well known in periglacial zone. Some lakes are of karst origin formed by the melting of permafrost. The lake water is rich in minerogenous compounds, especially in comparison with these in the lake water of western area. In general, bottom deposits in northern lakes are represented by peaty gyttja, in southern lakes by algal gyttja. The lakes of Galichkoe, Uchemerovo, Sindorskoe and Svetloyar have been studied (CHEBOTAREVA 1948; KOZLOVSKAYA 1956; KORDE 1956; SMIRNOVA 1981; TRUBE 1963).

2.8 The southern and eastern borders of the north-west lake district Boreo-nemoral zone (R- d) coincide with the limits of Weichselian glaciation. Whereas the lakes on the watershed of River Velikaya and Lovat (R- d_1) are mostly overgrown and their development has been touched upon in literature very briefly, lakes Seliger, Velikaya etc., on the Weichselian ice marginal zone (R- d_2) on the Heights of Bežanitski, Sudomi and southern part of Valdai have been investigated in greater detail.

2.9 The area, once covered by ice of Middle Pleistocene glaciations as well as the surroundings of periglacial area (R- e) are poor in lakes. At the same time these lakes are of the utmost importance while studying geological history of this area. The oldest deposits of L. Nero and Tatichevo are of Middle Pleistocene origin (ALESHINSKAYA 1974; ALESHINSKAYA & GUNOVA 1975). A 100 m thick core of terrigenous, carbonate, siliceous and deposits rich in organic matter have been discovered in the depression of L. Nero formed during Middle and Late Pleistocene and Holocene (ALESHINSKAYA 1974). During the Middle Pleistocene and Weichselian glacial epochs the oligotrophic and ultraoligotrophic conditions in L. Nero prevailed and terrigenous deposits were formed. During the Eemian interglacial and Holocene due to eutrophic conditions chemogenous and organogenous deposits accumulated (ALESHINSKAYA 1974). In L. Somino gyttja of remarkable thickness (40 m) has been discovered (NEUSHTADT 1936, 1965; KHOTINSKY et al. 1966). It has been subjected to detailed geological and palynological investigations in L. Glubokoe, Nerskoe, Beloe and Kochinskoe, nearby Moscow. These kettle basins are fulfilled with terrigenous and organogenous deposits which started to accumulate in Late-Glacial.

2.10 In woodland Steppe (R-g) there are
only a few lakes left on the terraces of
river valleys. Most of them are overgrown
already. Their stratigraphy, geology and
mollusc fauna have been studied very frag-
mentarily (Ilmen, Bobochnoe) without any
conclusions made about their deposits.

2.11 Steppe zone (R-h) is characterized
by very small number of lakes among which
the oxbow lakes dominated. Unfortunately
their history has not been studied yet.
The same can be said about Steppe-Desert
(R-i) lakes that are typical oxbow lakes
and located mainly in the river valleys or
delta plains. In some lakes the mineralization of lake water is so high that they
belong to the brackish water lake group.
Among these Elton and Baskunchac have been
studied more completely (SELIKHATOV 1933;
VASILYEV 1955; DZERES-LITOVSKII 1960).

3 REFERENCES

ALESHINSKAYA,Z.V., 1974: Paleogeogra-
phicheskye issledovanya v raione Roždest-
venskoi stoyanki na ozere Nero. - In:
Pervobytnyi chelovek, evo materialnaya
kultura i prirodnaya sreda v Pleistocene
i Holocene. IG AN SSSR, M. 279-283.

ALESHINSKAYA,Z.V. & V.S.GUNOVA, 1975: Holot-
senovaya istoria ozera Nero po dannym
sopryažennogo analiza. - In: Istoria ozer
v Holocene. Inst. Ozerovedenya AN SSSR,
L., 150-158.

BARTOSH,T.D., 1976: Geologia i resursy
presnovodnyh izvestkovistyh otloženii
holocena. Zinatne, Riga, 258 pp.

BERGLUND,B.E., 1979: Presentation of the
IGCP Project 158b. Paleohydrological
Changes in the Temperate Zone in the
Last 15 000 years - Lake and Mire Envi-
ronments. - Acta Universitatis Ouluensis,
82/A, 39-48.

CHEBOTARYEVA,N.S., 1948: Galichkaya ložbina.
- Uch. zap. Mosk. gor. ped. in-ta im.
V.P. Potyemkina 9, 7-42.

DZEREZ-LITOVSKII,A.I., 1960: Soljanye ozera
aridnoi zony zemnovo shara. - Tr. Lab.
ozerovedenya AN SSSR 10, 63-94.

ELINA,G.A., 1981: Printsipy i metody re-
konstruktsii i kartirovanya rastitel-
nosti holocena. - Nauka, L., 159 pp.

KHOTINSKII,N.A., A.L.DEVRIC & N.G.MARKOVA,
1966: Nekotorye cherty paleogeographii
i absolutnoi khronologii pozdnelednikovovo
vremeni centralnyh raionov Russkoi rav-
niny. - In: Verhnii pleistocen. Strati-
graphya i absolutnya khronologya. Nauka,
M., 140-151.

KORDE,N.V., 1956: Biostratigraphya otloženii ozera Uchemerovo i osnovnye etapy
evo istorii. - Tr. Lab. saprop. otl. 6,
83-109.

KOZLOVSKAYA,L.S., 1956: Istoria ozera Nero
po dannym izuchenya životnyh ostatkov. -
Tr. Lab. sapropelya 6, 173-180.

KVASOV,D.D. & E.S.MALYASOVA, 1969: Strati-
graphya donnyh otloženii nekotoryh ozer
Severo-Zapada po dannym sporovo-pyltse-
vovo analiza. - In: Ozera razlichnyh
lednikovyh landshaftov Severo-Zapada
SSSR, T. 2, Nauka, L. 136-138.

LAVRENKO,E.M. & V.B.SOCHAVA (ed.) Geobota-
nicheskaya karta SSSR, v mashtabe
1:400000 AN SSSR, M.-L., 1954.

MALYASOVA,E.S. & E.I.SPIRIDONOVA, 1965:
Novye dannye po stratigraphii i paleogeographii holocena Karelskovo pereshei-
ka. - In: Baltica, t. 2, Vilnius, 115-
124.

NEUSHTADT,M.I.,1936: K istorii pazvitya
ozer v poslelednikovoe vremya. - Pochvo-
vedenye 2, 269-276.

NEUSHTADT,M.I.,N.A.KHOTINSKII, A.L.DEVIRC
& N.G.MARKOVA, 1965: Ozero Somino (Yaro-
slavskya oblast). - In: Paleogeographya
i khronologya verkhnevo pleistocena i
holocena po dannym radiouglerodnovo me-
toda. Nauka, M., 91-97.

SEMENOVICH.N.I.,1958: Limnologicheskye us-
lovya nakoplenya železistyh osadkov v
ozerah. AN SSSR, M.-L., 188 pp.

SELIKHATOV,A.I.,1933: Materyaly dlya geo-
logii Baskunchatskovo ozera. - Tr. Vses.
geologorazvedoch. obedin. NKTP SSSR, vyp.
284, 3-24.

SJÖRS,H., 1963: Ampi-Atlantic zonation,
Nemoral to Arctic. Löve,A. & Löve,A.(eds.).
- North Atlantic Biota and their History.
Oxford, 109-125.

SMIRNOVA,V.I., 1981: Istorya ozera Sindors-
kovo (Komi ASSR) v Holocene (po dannym
diatomovovo analyza. - In: Sistematika,
evolucya i ekologya vodoroslei i ih zna-
chenie v praktike geologicheskih issle-
dovanii. Kiev, 136-137.

TRUBE,L.L., 1963: O proiskhoždenii oz.
Svetloyar v Gorkovskoi oblasti. - Izv.
Vses. Geogr. ob-va, t. 95, vyp. 5, 454-
455.

VASILYEV,G.A., 1955: Chetvertichnye otlo-
ženya ozera Elton i istorya ih obrazo-
vanya. - Tr. VNII galurgii, 30, 205-223.

Lake, Mire and River Environments, Lang & Schlüchter (eds)
© 1988 Balkema, Rotterdam. ISBN 90 6191 849 9

Synchronous pollen changes and traditional land use in south Finland, studied from three adjacent sites: A lake, a bog and a forest soil

K. Tolonen
University of Oulu, Oulu, Finland

M. Tolonen
University of Helsinki, Helsinki, Finland

ABSTRACT: Previous diatom analyses on a sediment core spanning the late Holocene from a remote forest lake (Lake Vitsjön; 59°58'N, 23°19'E) revealed changes that were attributed possibly to early land use around the basin. Fine resolution pollen analysis was undertaken on the uppermost 95 cm of a replicate short core from the lake. The pollen was compared with that from a peat profile from an adjacent small bog and that from a forest soil. The lake core and the peat core were dated by a total of six radiocarbon dates. When using the same basic AP pollen sum from all sites and types of deposits, the history of forests and land use in the study area was evident most clearly in the pollen profile from the lake. The onset of rye cultivation in the region was radiocarbon dated to the Viking period, about A.D. 800, in the Vitsjön core. However, it was not until the hydroseral development of the mire had successed from an alder and birch swamp to a more open Sphagnum-pine bog in early Medieval Times (after A.D. 1200) that any signs of human activity became visible in the pollen spectra from the peat core.

Carbonized seeds of Hordeum vulgare with other macrofossils of weeds in the soil core provided direct evidence for ancient in situ cultivation in the vicinity of the lake.

1 INTRODUCTION

Pollen studies from open sites, such as lakes and some peat bogs are recommended as primary reference sites for the description of past regional vegetation history (BERGLUND & DIGERFELDT 1976, BERGLUND 1979). It has been recommended also, that complementary pollen information ought to be obtained from small hollows with peat, gyttja deposits, or soil profiles with deep humus layers in some areas. In fact, such small basins have proved to be very valuable because their pollen source area is often very restricted. Consequently in conjunction with regional lake pollen profiles, they contribute to detailed reconstructions of the history of man and the landscape (IVERSEN 1964, 1969, ANDERSEN 1973, 1979, AABY 1983).

Lake Vitsjön and its immediate vicinity, representing a relatively remote area in Tenala (Tenhola), southwestern Finland, was chosen to compare radiocarbon dates and fossil pollen records among the sites, and then to reconstruct past traditional agriculture practices including slash-and-burn cultivation and pasturing in the region.

The initiative for this study arose from unexpected stratigraphic changes in diatoms in the sediments of about the last 3000 years (Paleolimnologian kurssi 1979, K.TOLONEN et al. 1986). The reconstructed history implied a long-term decrease in lake pH following isolation of the basin from the Baltic Sea (at about 2.1 m) that was subsequently interrupted by a temporary and very distinct pH rise at a sediment depth between 0.7-0.3 m below the present water/mud interface. This was peculiar because the lake shore presently lacks fields, settlement or any other signs of human disturbances, and the lake is situated in a region of dense coniferous forest that is sparsely populated. The closest farms are 0.5 to 1 kilometer away. In addition, the catchment of the lake is restricted to a very narrow belt around the lake basin (Fig.1).

Preliminary pollen analysis carried out in 1979 (Fig.4) revealed a number of rye pollen grains and many other "human indicators" such as Rumex acetosella and Juniperus communis.

The present study was undertaken to provide a more detailed paleoecological basis

Table 1. Some water parameters from Lake Vitsjön (59°58'N, 23°19'E; elevation 16.3 m a.s.l.). The lake area is c. 28.5 ha and the drainage area reaches 88 ha. The data originate from one sampling carried out in connection with the field course in limnology. Dept. Limnol., Univ. Helsinki (LEHMUSLUOTO 1976, unpublished) and from two samplings carried out in May 1978 and October 1979 during the spring and autumn overturn by K.Tolonen.

Depth m	1 – 5	7 – 9	11
Temp. °C	20.5 – 20.2	14.6 – 8.7	7.2
5.5.1978	6.0	–	5.4
20.10.1979	0.8	6.8	6.8
pH	5.9 – 6.1	5.8 – 5.5	5.6
5.5.1978	4.8	–	4.9
20.10.1979	5.2	–	5.3
Oxygen mg/l	8.4 – 8.3	9.7 – 31	0.0
Cond. $\mu S_{20°C} cm^{-2}$	40 – 39	39 – 42	44
Colour mg Pt/l	17 – 15	26 – 34	114
Ptot · mg/l	0.005 – 0.007	0.011	0.023
SBV mg l^{-1}	0.05	0.06	0.2
NO_3-N mg l^{-1}	0.01	0.01	0.00
Fe mg l^{-1}	<0.1	0.1 – 0.2	1.4

to identify the possible areas of agricultural activity that were responsible for the cultural pollen types present in the lake sediment. Two additional study sites, one a peatland site and the other a forest soil profile were chosen for comparison with the results from the lake, and to assess the reliability of each type of deposit for radiocarbon dating and paleoecological reconstruction.

2 THE STUDY SITES AND REGIONAL SETTING

Lake Vitsjön is a small oligotrophic forest lake (Table 1) with an undisturbed catchment area (Fig.1). The most common vascular plant species growing in the lake and along the shore include Nuphar lutea, Nymphaea candida, Lobelia dortmanna, Isoëtes lacustris, Ranunculus reptans, Juncus bulbosus and Typha latifolia.

The maximum depth of the lake is about 12.3 m. The bottom most 1 m layer of water in deeper parts of the basin can experience an oxygen deficit during the late summer and late winter. The surface sediment, however, is not laminated and is dark brown in colour. The water chemistry of the lake has been monitored for many years by BECKER (1978), LEHMUSLUOTO (Dept. of Limnology, University of Helsinki, unpubl.), the General Board of Waters (KENTTÄMIES unpubl.) and the authors (unpubl.). Accorsing to these data, the pH of the surface water layer is about 6.0. The buffering capacity of the lake is so low that during the autumnal overturn of 1976 (on Oc-

tober 10) the pH of the lake was 5.2 in the uppermost 1-7 m, and the pH was 5.3 for the 9-11 m layers (LEHMUSLUOTO, unpubl.). Even lower values were recorded during snowmelt on May 5, 1977: pH 4.8-4.9 for the water of 1-11 m (BECKER 1978).

The second site, a small (about 80x300 m) forest-covered peatland, is located about 320 m north of the lake. Pine (Pinus sylvestris) and spruce (Picea abies) are the dominant species on the peatland, accompanied by scattered birch (Betula pubescens). Dwarf shrubs, such as Calluna vulgaris, Empetrum nigrum, Vaccinium uliginosum, V.myrtillus, V.vitis-idaea, V.oxycoccos and Ledum palustre occupy the field layer together with Eriophorum vaginatum and Carex globularis. The moss layer is dominated by Sphagnum angustifolium, S.magellanicum, Polytrichum commune and Pleurozium schreberi. It is probably best classified as an ombrotrophic peatland (bog) since there are no minerotrophic species in the mire, except for Betula and Picea.

The last site is a soil profile that was sampled from a small shallow depression. It is covered by Ledum palustre, Carex globularis, Calluna vulgaris, Polytrichum commune, Sphagnum girgensohnii, S.russovii, S.angustifolium and relative young deciduous trees (Betula, Salix etc.). The site is located about 160 m north of the lake (Figs. 1 and 2). There are numerous charred tree stumps in the depression which indicate that the undulating surface of the site once had been covered by mature forest, despite that today it is semi-open.

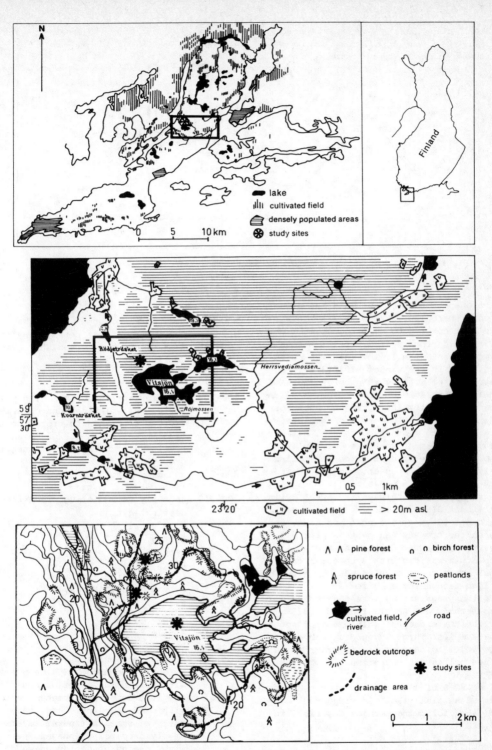

Figure 1. Maps showing the location of the study area in the central part of Hanko Peninsula. The coring sites are denoted with asteriks.

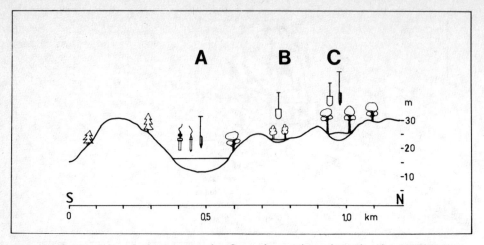

Figure 2. Profile of the topography from the coring sites in the study area.

3 METHODS

The deepest part of the lake basin was cored on March 20, 1979 with a large Russian peat corer (100 cm long, 10 cm wide) and a "cold finger" iron tube (3 m long, 6 cm in inner diam.). The lake was cored once again in the winter of 1983 with a pistonless plexiglass tube corer (ZÜLLIG model) having a 10 cm inner diam. A 95 cm long core was obtained from the uppermost sediment strata. The core was sectioned into 95 slices (each 1 cm thick). A small volume of sediment was taken carefully from the centre of each slice for pollen analysis during the sectioning of the core with an adjustable pipet.

A sharpened and thin walled steel cylinder (inner diam. 11 cm) was cut through the upper 60 cm at the peat bog site, and thereafter the Russian peat sampler was used to core lower depths. At the soil site a pit was dug with a spade. A continuous sequence of samples was collected on a clean face using a knife.

Pollen preparations were made using standard methods of FAEGRI & IVERSEN (1975), however KOH treatment only was used on the lake samples. Three Lycopodium tablets were added to each sample in order to calculate pollen concentrations (STOCKMARR 1971).

The ^{14}C Dating Laboratory at the University of Helsinki (Hel) provided radiocarbon dates following methods described in JUNGNER (1979).

4 RESULTS

4.1 Description of sediment and soil strata of the three sites:

a. The lake

cm
0-8	very dark coloured mud
8-90	dark green coarse detrital gyttja
90-138	dark brownish to greenish coarse detrital gyttja
138-160	greyish green coarse detrital gyttja
160-208	olive green fine gyttja with yellowish-white laminations clearly visible from in situ frozen cores (isolation niveau at 205 cm as concluded from the diatom stratigraphy)
208-209	annually laminated purple red fine detrital gyttja
209-219	annually varved olive green fine detrital gyttja
219-230	fine detrital gyttja or clay-gyttja
230-290	greenish grey clay gyttja with black sulfide striations
290-300	black (sulfide) streaked clay-gyttja or gyttja-clay
300-	silty fine sand or sand

b. The bog

cm
0-20	unhumified Sphagnum peat
20-35	slightly decomposed Sphagnum peat with remains of dwarf shrubs
35-55	highly decomposed Sphagnum peat with woody remains. Dark, very humified streaks (?burnt layers)
55-80	highly decomposed Sphagnum peat with remains of deciduous wood
80-90	very compact detrital gyttja, black to dark grey, the upper and

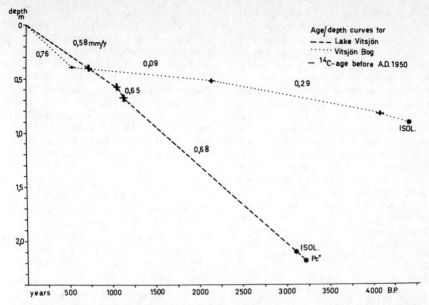

Figure 3. Age/depth curves for Lake Vitsjön long core (L) and Vitsjön Bog (B) based on conventional radiocarbon ages (single S.D. given with bars), on known ages for isolation niveau (Prof. G.Gluckert, pers, communication May 1985), as concluded from the diatom stratigraphy. Also previous radiocarbon dating for the general spread of Picea (Pc+) in the area was used.

lower boundary diffuse
90-110   clay gyttja, light grey. Upper boundary diffuse (isolation niveau between 90 and 95 cm as concluded from the diatom stratigraphy)

110-   silty clay, grey. Upper boundary diffuse.

c. The soil profile

cm
0-2   F   fresh organic material with Pleurozium schreberi, roots of Vaccinium vitis-idaea

2-13   H   weakly decomposed forest peat, dark brown, rich in root remains and Sphagnum

13-17   H   moderately decomposed forest peat, reddish brown

17-20   Ah   highly decomposed organic material, dark brown to black, some charcoal fragments

20-32   Ah   dark brown silt containing horizon with abundant humus and layers of charcoal fragments, soft

32-37   B   coarse sand
37-   C   fine sand

4.2 Radiocarbon dates and sediment accumulation

In all, six radiocarbon dates were obtained from the lake and bog profiles. These are as follows:

Site	Depth,cm	Lab.number	Radiocarbon years B.P.
Lake	38 – 44	Hel-1984	720 ± 140
"	56 – 62	Hel-1985	1060 ± 140
"	66 – 72	Hel-1986	1170 ± 150
Bog	40 – 42	Hel-1987	550 ± 90
"	54 – 58	Hel-1988	2180 ± 90
"	84 – 88	Hel-1989	4170 ± 110

All the dates are stratigraphically in order and are in agreement with the regional pollen chronology. A consistent picture emerges for sedimentation rates by combining the radiocarbon dates with the results of the pollen analyses and the known shoreline displacement for the area (the isolation niveau being studied from the diatoms). It can be supposed that during sediment accumulation after isolation of the basin, there have been no major disturbances, and thus, no abrupt changes in the vertical growth of the sediment.

If the radiocarbon dates are fitted on age-depth curves, sediment deposition in the lake appears to have been relatively uniform, i.e. about 0.5-0.7 mm/yr. (Fig.3).

At the base of the profile for the humified carr peat sediment deposition was about 0.29 mm/yr. Between 40 cm and 55 cm depth the deposition rate was as low as 0.09 mm/yr, and in the upper 40 cm a rate of about 0.76 mm/yr in the Sphagnum peat was maintained. From the stratigraphy, we find that peat between 40-55 cm depth (Fig. 6) was highly humified with dark streaks, obviously denoting burnt layers. It becomes apparent that in situ fires contributed to spurious sediment accumulation rates. The upper 40 cm consists of raw and poorly decomposed peat, and the accumulation rate seems reasonable in the light of general peat growth rates in southern Finland.

4.3 Zonation of pollen diagrams

The pollen diagrams of the peat core and the two lake cores (Fig.4-7) are divided into local pollen zones, that are attributed to various successional stages in forest development. The zones are numbered from the base upwards, and are prefixed with the site initials, LV or VB. The main pollen characteristics of the zones will be discussed, and interpreted in terms of the vegetation immediately surrounding the site.

4.3.1 Lake Vitsjön

LV-1.a. (210-230 cm)
This is the lowermost subzone, where Betula, Alnus and QM together comprise up to 75% of total arboreal pollen. Herbaceous pollen also occur in high frequencies. Picea pollen rises near the top of the subzone.

LV-1.b. (180-210 cm)
The lower boundary is placed at the increase of Alnus (A.glutinosa). The isolation of the basin from the Baltic Sea is at 205-210 cm. There is a rise in Salix and Juniperus associated with occurrences of Myrica, Thalictrum and Poaceae. This suggests that deciduous shore vegetation developed relatively rapidly along the receding lake shore.

LV-2 (130-180 cm)
This zone is defined by a relatively steady Betula pollen curve. The Pinus and increasing Picea curves are predominant. This suggests a primary successional process that lasted for about 1000 years.

LV-3.a. (50-130 cm)
In this subzone several phases of inter-related changes in arboreal pollen percentages occur, and the contribution of QM pollen falls throughout. In the upper part of the subzone herb pollen values increase. Values of Juniperus, Poaceae, Rumex acetosa are relatively high and a few spores of Isoetes are present. Correspondingly with the increase of the number of herb pollen taxa, there is Cereal pollen type (including Secale) from 74 cm level.

LV-3.b. (15-50 cm
Pinus and Juniperus frequencies are progressively increasing while Betula and Picea values fall. There is a corresponding rise in values of herb pollen, due mainly to Poaceae, Rumex and cereals. Pollen of Plantago ssp., Centaurea cyanus and Epilobium are present.

LV-3.c. (0-15 cm)
This subzone commences with a clear decline in Cerealia pollen type coupled with the fall of Juniperus.

4.3.2 Vitsjön Bog

VB-1.a. (95-130 cm)
High, but decreasing values of broad-leaved deciduous trees (Corylus, Ulmus, Quercus) accompanied by Betula and Pinus, and some Poaceae, Filipendula and Polypodiaceae typify this subzone. The sediment is highly minerogenic (Fig.6). The basin was flooded by the sea at this time. It is possible, that part is redeposited from pollen contained in the sea water. Herbs of open habitat, such as Chenopodiaceae, Artemisia and Rumex acetosa are present in notable abundance, suggesting that disturbed soils and open conditions occurred in adjacent areas bordering the Baltic Sea. Pollen of Nymphaea indicates open water.

VB-1.b. (85-95 cm)
There is a gradual fall in the QM element. At first various herbaceous plants show an increase, followed by Alnus which predominates among the tree taxa. Salix and Cyperaceae expand abruptly at the close of the zone. The reason for this was due to the isolation of the basin from the sea accompanied by a relatively rapid development of deciduous shore vegetation around the pool.

VB-2 (60-85 cm)
The peat-forming community, indicated by the pollen diagram, was dominated by Sphagnum mixed with carr herbs such as

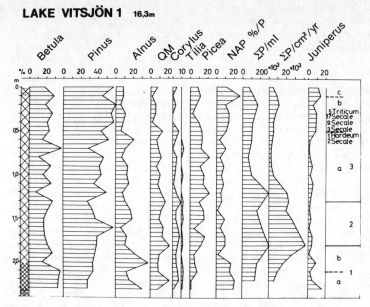

LAKE VITSJÖN 1 16,3m

Figure 4. Orientative pollen diagram for Lake Vitsjön,
site A, prepared by the course in palaeolimnology in 1979
(Palaeolimnology kurssi 1979).

Lysimachia and Cornus. In this zone Alnus
is replaced by Betula, suggesting that
successional processes were progressing
towards the establishement of closed birch
forest on the mire. Poaceae and Cyperaceae
continue to be the most frequent herbaceous plants, suggesting local marsh or
swamp conditions.

VB-3.a. (35-60 cm)

Overall in this subzone, there is a rise
in the proportion of herbaceous pollen.
Pinus predominates among the trees while
Betula falls. There is a rise in Picea
pollen, which suggests that the forest
cover of the bog became more open expanding the pollen source area around the site.
The general migration of Picea to the area
took place around the same time as the
isolation of the lake basin (ca. 3200 B.P.)
from the sea, which is documented in the
profile from Lake Vitsjön also (Fig.4).
There was a slow sedimentation rate in
this section of the core. Correspondingly,
several levels contained abundant macro-
scopic and microscopic charcoal fragments.
There is a rise in Calluna and Ericaceae
pollen. These changes suggest the formation
of bog vegetation. The occurrence of Cerealia type pollen grains (including Secale), associated with the more regular
presence of Rumex acetosa type and Artemisia from the 41-40 cm level suggests

that agriculture was practised in the vicinity. Burnt layers in the stratigraphy
between 50-40 cm and numerous Epilobium
pollen at various depths thus may indicate
forest clearance which can be linked with
cereal pollen in the diagram. At the same
time there is a fall in the Picea curve.

VB-3.b. (15-35 cm)

In this subzone there is a brief rise
first in Alnus followed by Betula, and
after that a maximum in herbaceous pollen.
These trends are suggestive of a renewed
forest clearance. Marked peaks in cereal
pollen associated with various open habitat and weed species (Plantago lanceolata,
Rumex acetosa type, Chenopodiaceae and
Artemisia) indicate intensive cultivation
activity regionally or activity close to
study site. Among the herbs and dwarf-
shrubs Rubus chamaemorus, Drosera and
Ericaceae appear, probably reflecting local communities on the bog. Degraded and
broken pollen is high at the 30 cm level.
Also there is a high content of charcoal
fragments in the peat. These changes associated with an inwash of mineral particles
into the basin, suggests the maintenance
of disturbed soils around the bog.

VB-3.c. (0-15 cm)

There follows in the upper 15 cm of the
profile, an overall rise in the proportion

89

of tree pollen. Pinus predominates among the trees. This phase suggests the abandonment of agricultural activity and reforestation of the area.

4.3.3 Soil site

The pollen profile in the present study cannot be compared directly with the lake core and the peat core due to complex processes associated with development of the soil profile (e.g. mixing of soil, differential destruction and deterioration of pollen and downwash of pollen).

The changes in the pollen profile may be asynchronous with those in the other profiles. The pollen content is low below 21 cm level. That part of the profile is interpreted to represent an in situ clearance and cultivation horizon, as concluded from the charred seeds of Hordeum vulgare, Galium sp., Vicia sp. and Fallopia convolvulus, found at the charcoal rich layer between 21 and 32 cm (Fig.8).

At 15-20 cm the proportion of pollen of the different taxa are not considered very reliable, as the pollen assemblages may have mixed through downwash or turning of soil.

In the upper 15 cm the raw peat of the profile supposedly represents primary deposition and therefore the pollen stratigraphy and the changes in the proportions of the arboreal species are considered to reflect real changes in the vegetation. The changes are referred to as local in the light of general studies on pollen dispersal within forests (ANDERSEN 1970).

The cerealia pollen curve between 5 cm and 21 cm is rather constant, and nearer the surface it is not present at all. The presence of cerealia pollen (including Secale) suggests that the accumulation and formation of the deposit has taken place during the agricultural period. The occurrence of Centaurea cyanus pollen at several levels suggests open ground close to the site associated with cultivation of winter cereals.

The pollen composition of the surface sample (0-1 cm) is slightly different from the spectra of the lower sediments (Fig. 7). It is dominated by high Betula (>73%). The pollen of Calluna and Poaceae closely reflects the present vegetation around the depression. Trees have overgrown the site, and it is of interest that in the surface sample, there is no Cerealia type pollen present, though the surface sample of the lake core revealed Cerealia ca 1% of the total pollen sum. The nearest fields today are situated to NE some 600 m from the coring site.

4.4 Comparison of pollen stratigraphies

The interpretation primarily concerns the development of the local terrestrial vegetation. The lake vegetation is described only briefly in the diagram since the results show no major changes in the macrophyte vegetation during the last 3000 years. The dynamic and progressive primary succession which characterizes the vegetational development during the period in question was controlled by pedogenic factors along the receding seashore as well as the competition of the different species. The pollen diagrams made from Tenhola resemble other diagrams from SW Finland, and the local pollen assemblage zones can be assumed to have a partly regional application (K.TOLONEN et al. 1979, M.TOLONEN 1983, 1985).

Generally it can be concluded that although the local forest pollen zones are synchronous, and showing the same developmental pattern, it has been possible to demonstrate that there are differences between sites in both temporal and spatial variation of pollen frequencies. The tree pollen curves in the lake profile are more constant in the long-term run than are those in the bog profile. The distinct changes in the bog profile reflect more local pollen rain and the over-representation of pollen from Betula at earlier successional period and later Pinus is obvious. The pollen frequencies in the lake profile throughout may show more regional abundances of the tree componets in the forests, but the curves likely are smoothened as a consequence of depositional processes in the lake. Picea and QM, if related to their real proportion in forest, are not that much underrepresented in the lake profile as they are in the peat profile.

From the beginning of Cerealia pollen curve in the LV profile, accompanied with increasing frequencies of various NAP (Juniperus, Rumex and Epilobium), it appears that human activity and forest clearance in the area started at 1170 ± 150 B.P., corresponding to the turn of the Merovingian time to the Viking Period. It is difficult to conclude the effective source area (TAUBER 1965) of the lake pollen diagram. DIGERFELDT (1982) concluded, based on the area of his study lake (ca. 25 ha), that the pollen diagramm primarily records

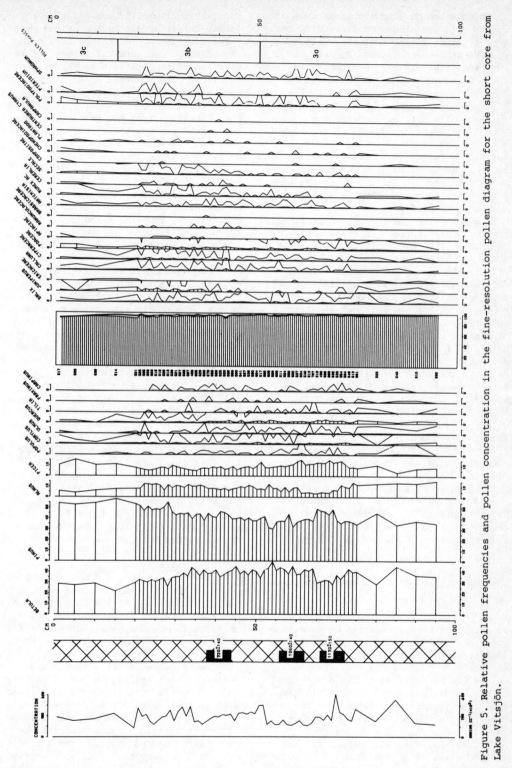

Figure 5. Relative pollen frequencies and pollen concentration in the fine-resolution pollen diagram for the short core from Lake Vitsjön.

91

the vegetation development in an area of 10-15 km radius around the lake. In the present study the area of Lake Vitsjön is 28.5 ha (the drainage area 88 ha), and correspondingly it can be speculated that the pollen source area can be an area of some 10-15 km radius around the lake.

In the VB diagram the beginning of Cerealia pollen type as a continuous curve with a contemporaneous well-marked rise in weed species is dated to 550 + 90 B.P. This horizon is concluded to be synchronous with an increase in signs from human influence above the date 720 + 140 B.P. at LV; the percentages found for cereals are equally high in both profiles.

In common with the bog pollen profiles, the long-term decrease in Picea pollen, from about 1100 B.P. until 250 B.P. is clear at the lake pollen diagram. Otherwise in the latter case the short-term successional changes are not significantly interpretable. Since the pollen diagrams were made with close interval sampling, the difference may be attributable to the dissimilar pollen deposition and sedimentation in the lake and bog basins. One of the major factors explaining this difference is the effective filtration of certain pollen by the dense (deciduous) forest on the mire site (cf. K.TOLONEN et al. 1979, M.TOLONEN 1985).

Increases in grass, heather, alder birch and pine, respectively, indicate short-term vegetation succession of post-fire communities on and around the bog at 550-250 B.P.

The renewed expansion of the forest, continued from ca. 200 B.P. until the present day is indicated in all diagrams by an expansion of Pinus forest and by distinct decrease in open ground herbs and Cerealia pollen.

5 RECONSTRUCTION OF ENVIRONMENT AND LAND-USE

The pollen diagrams give evidence of long-term primary succession of vegetation starting from some 4500 years B.P. in the area of land uplift in the archipelago of the northern Gulf of Finland. A series of stages can be outlined in the development of the woods, but their spatial distribution cannot exactly be delimited. The peat formation in the bog site started some 800-900 years after the isolation of the basin. By that time (ca. 3600 B.P.) the sea shore was at 20 m contour; i.e. between the present lake and the bog. The vegeta-

tion was unstable. By about 3200 B.P. spruce forest spread to the upland areas.

The distribution of the primary broad-leaved deciduous forest followed the regression of the Baltic Sea. The evidence available here suggests that the regional vegetational cover did not undergo any significant change or major opening before the transition from the Merovigian time to the Viking period. Until ca. 800 B.P. the mire and its adjacent surroundings were carrying a dense deciduous wood.

The results obtained show, that from about 1100-1200 B.P., however, temporary forest clearances with cereal cultivation on edaphically favourable areas of the geologically young mainland, ought also be included as a factor in the vegetation succession. From ca. 750 B.P. the mire basin changed to a more open pine bog either as a consequence of a natural long-term edaphic and hydrologic change, or accelerated by human activity or both. In consequence, a more diverse pollen rain was able to enter the bog site. By burning patches of forest the distribution pattern of the age class of the forest stands changed continuously. The bog might also have been periodically totally treeless as concluded from the burnt peat horizons in the profile. The correspondingly high frequencies of Juniperus in the LB profile must originate from open hilly upland areas around the lake.

The frequencies of Picea show a long-term decrease, starting at around 1000 years B.P.. It probably indicates regional reduction in production of spruce pollen. The renewed expansion of the forest, which has continued from ca. 200 B.P. until the present day, is indicated in the diagram by an expansion of Pinus forest and by a distinct decrease in open ground herbs and Cerealia pollen.

In the present context, macrofossil evidence of the cultural communities preserved at the mineral soil site suggests evidence on local cultivation. From the soil profile, it would appear that local clearance and cultivation was made by burning birch-alder forest. Here it is proposed that the lowest sequence (A) with its high pollen content and prominence of indicators of cereal cultivation is synchronous with the beginning of the Cerealia pollen in the VB profile. This comparison is not without limitations, since there is no radiocarbon dating made for the soil site. As pollen preservation declined rapidly with depth in the soil profile reconstruc-

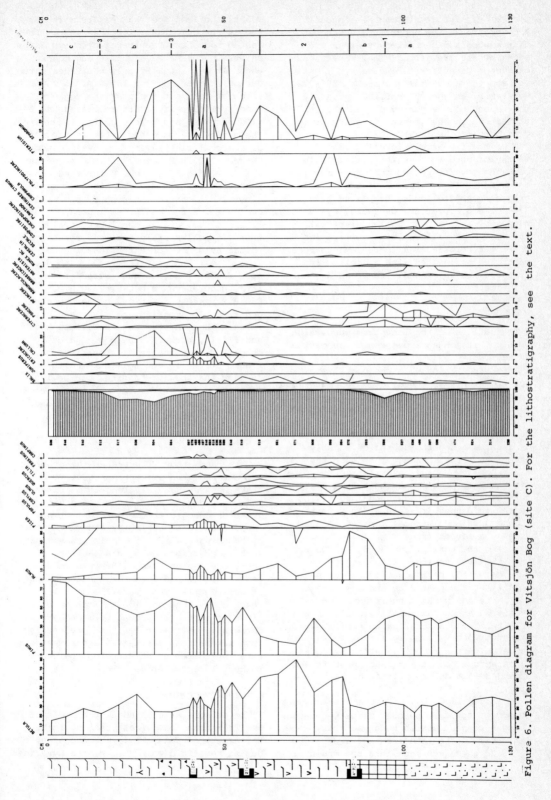

Figure 6. Pollen diagram for Vitsjön Bog (site C). For the lithostratigraphy, see the text.

tion of earlier development is impossible. The spectra at the upper levels suggest that a woodland dominated by Pinus and Picea became prevailing. The simultaneous maxima of Cerealia (including rye), Calluna, Poaceae and Rumex acetosa at 250-375 years B.P. in the VB profile, certainly correspond to the period of maximal land-use either in the general vicinity or in this part of the area.

From the fine-scale local variation worthy of note is that the soil profile contains several pollen grains of Centaurea cyanus. It is also present at levels preceding and following the dated level, 720 ± 140 B.P., at LV. Centaurea cyanus supposedly has a poor pollen dispersal. This seems to reflect local cultivation of winter cereals. The increase in charcoal particles assiciated with a decline in Picea pollen is linked with slash and burn practice.

6 SUMMARY

The outline of the long-term succession of Lake Vitsjön environs can be summarized as follows:

After the period when the bog site emerged from the Baltic Sea, at about 4400-4500 years ago, until the isolation of the lake basin at 3100 B.P., deciduous woodland progressively replaced the primary shore meadow vegetation that was open. The number of grasses and herbs decreased shortly after the forest became closed. At the same time when new boulder skerries kept appearing in this part of the then Baltic archipelago, the bog site changed into drying carr with dense birch and alder forest. At around 3200 B.P. spruce migrated to the area as a forest tree.

During the period from 3000 to 2000 B.P. when the surroundings of the study area were mainland, deciduous forest dominated locally at the peat site and the forest soil site and perhaps their adjacent water receiving areas. Regionally again, in the long-term successional pattern of vegetation, developing in a series of belts parallel with the receding Baltic Sea coast, first spruce forest and later pine forest expanded into drier soils at higher elevation.

At around 1500-1000 B.P. the mire changed to ombrotrophic pine bog. First vegetation changes connected with human activity were indicated in the Late Iron Age (1200-1000 B.P. and these continue into the Medieval

Times. The Iron Age changes are more marked at the open lake site than in the forested bog site; the latter displays major shifts in the Medieval times and modern ages. It was actually only the Picea curve during the period from 100 B.P. to 250 B.P. which at both sites showed the regional long-term decrease. This indicated the period of slash and burn practice in the area. The evidence at the bog site, in the form of charcoal rich layers at levels dated between 1200 B.P. and 375 B.P., suggests that periodically it was unforested and respectively more diverse plant communities with short-term secondary successions were born. Fires indicated increased waterlogging, too, and at the forest soil site, there is pollen analytical and stratigraphical evidence for extension of waterlogged soil and peat referred to above. In more recent time, in the course of some three hundred years between 550 B.P. and 250 B.P., the vegetation on the bog changed from pine moorland to alder and birch forested bog, after that to Calluna heath and finally back to pine bog, again. It is suggested that the in situ cultivation, traced at the forest soil site, although lacking the absolute dating, can be linked with the major shifts in vegetation recorded in the lake and bog pollen diagrams during the period under consideration.

7 CONCLUSIONS

This case study illustrates the problems encountered when attempting to differentiate between natural (long-term primary) and man-induced (short-term secondary) vegetation changes in the development of landscape. This particular study shows that by choosing critical-sites in study areas, such as described here in adjacent environs, one can observe evidence of concurrent regional and local vegetation changes in some detail. The overall conclusion reached, is that the identification of periods of instability must be made by a correlation of several sites, preferably from lake sites, peat sites and forest soil sites which represent different paleoenvironmental repositories.

8 ACKNOWLEDGEMENTS

The authors wish to thank the staff of the Tvärminne Zoological Station of the University of Helsinki for technical help during the field work. Ms. Riitta Hyvärinen

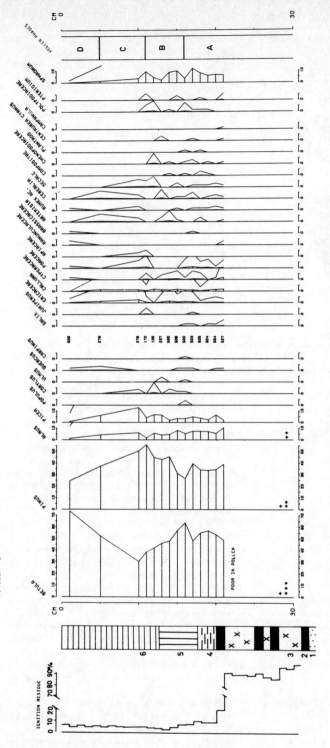

Figure 7. Pollen diagram for the Vitsjön soil profile (site B). Lithostratigraphy given in the text.

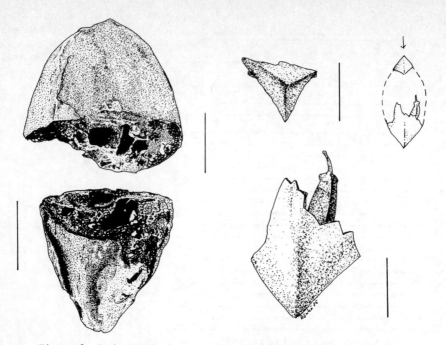

Figure 8. Carbonized plant remains encountered from Vitsjön
soil profile (site B), depth 21-32 cm. 1. Hulled barley(Hordeum
vulgare). 2. Wild buckwheat (Fallopia convolvulus), Scale 1 mm.

is cordially acknowledged for pollen analyses, Dr. Barry Warner for revising the
language and Ms. Pirkko Dookie for type-
writing the text.

9 REFERENCES

AABY,B., 1983: Forest development, soil
genesis and human activity illustrated
by pollen and hypha analysis of two
neighbouring podzols in Draved Forest,
Denmark. - Danmarks Geologiska Under-
søgelse, II. Raekke 114, 114 pp.+Appen-
dix 1 & 2.

ANDERSEN,S.T., 1973: The differential
pollen productivity of trees and its
significance for the interpretation of
a pollen diagram from a forested region.
- In: BIRKS,H.J.B. & R.G.WEST (eds.),
Quaternary plant ecology, 14th Symp.
Brit. Ecol. Soc., Univ. of Cambridge 1,
109-115.

ANDERSEN,S.T., 1979: Brown earth and podzol: Soil genesis illuminated by micro-
fossil analysis. - Boreas 8, 59-73.

BECKER,P., 1978: Levätestien tulostusta-
voista ja järvien talvisesta tilasta
levätestien ja kemiallisten analyysien
valossa. - M. Sci. Thesis. Dept of Limnology, University of Helsinki, 150 pp.
(mimeogr.).

BERGLUND,B.E. (ed.), 1979: Palaeohydrolo-
gical changes in the temperate zone in
the last 15 000 years. Subproject B.
Lake and mire environments. - Project
Guice, Vol. I. International Geological
Correlation Programme, Project 158. Lund.
LUNBDS/(NBGK-3001)/1-140(1979).

BERGLUND,B.E. & G.DIGERFELDT, 1976: Environmental changes during the Holocene - a
geological correlation project on a Nor-
dic basis. - Newsl. Stratigr. 5(1): 80-
85 + 3 Figs., Berlin.

DIGERFELDT,G., 1982: The Holocene development of Lake Sämbosjön. 1. The regional
vegetation history. - Univ. of Lund,
Dept. of Quaternary geology, Report 23,
1-24. Coden: LUNBDS/(NBGK-7023)/1-24/
(1982).

FAEGRI,K. & J.IVERSEN, 1975: Textbook of
pollen analysis. - 3rd ed. Munksgaard,
295 pp.

IVERSEN,J., 1964: Retrogressive vegetational succession in the Post-glacial. -
J.Ecol. 52 (Suppl.): 59-70.

IVERSEN,J., 1969: Retrogressive development of a forest ecosystem demonstrated
by pollen diagrams from fossil mor. -

Oikos, Suppl. 12, 35-49.

JUNGNER,H., 1979: Radiocarbon dates. I. - Radiocarbon Dating Laboratory, Univ. Helsinki, Report 1: 1-131.

PALAEOLIMNOLOGIAN kurssi 1979: Tvärminne. Course in Palaeolimnology 1979. Tvärminne 19.-30.3. Research Results, Tvärminne Zoological Station, University of Helsinki, 20 pp (mineogr.).

STOCKMARR,J., 1971: Tables with spores used in absolute pollen analysis. - Pollen et Spores 4, 615-621.

TAUBER,H., 1965: Differential pollen dispersion and the interpretation of pollen diagrams. - Dan. geol. unders. 2. rk. 89.

TOLONEN,K., M.LIUKKONEN, R.HARJULA & A.PÄTILÄ, 1986: Acidification of small lakes in Finland studied by means of sedimentary diatom and chrysophycean remains. - Hydrobiologia (in press).

TOLONEN,M., 1983: Late Holocene vegetational history in Salo, Pukkila, SW Finland, with particular reference to human interference. - Ann. Bot. Fennici 20, 157-168.

TOLONEN,M., 1985: Palaeoecological reconstruction of vegetation in a prehistoric settlement area, Salo, SW Finland. - Ann. Bot. Fennici 22, 101-116.

Lake, Mire and River Environments, Lang & Schlüchter (eds)
© 1988 Balkema, Rotterdam. ISBN 90 6191 849 9

Sedimentation and local vegetation development of a reference site in southwestern Bulgaria

S.Tonkov
Biological Faculty, University of Sofia, Bulgaria

ABSTRACT: Pollen-analytical and stratigraphical investigations were carried out in Southwestern Bulgaria on two cores from the largest minerotrophic topogenous fen in respect to study its formation and the vegetation history. Several stages in the development of the basin and the local vegetation during the last 8000 years were delimited on the basis of a detailed description of the sediments accumulated and the changes in the pollen record supported by 9 radiocarbon datings.

1 INTRODUCTION and METHODS

In the frame of the national IGCP-Project 1958 B programme a complex palaeoecological research of the largest inland fen Tschokljovo marsh was carried out. The site investigated is located at 870 m a.s.l. in the central part of Konjavska mountain (1487 m) and occupies a territory of 1.8 km^2. During the last twenty years this place has been used for peat cutting and was completely drained which had a negative effect on the natural marsh vegetation. This fen had been famous for the occurrence of some rare plant species in the bulgarian flora such as Salix pentandra L., S.rosmarinifolia L., Peucedanum palustre (L.)Moench., Pedicularis palustris L., Potentilla palustris (L.)Scop., Thelypteris palustris Schott. etc.

The basic analytical work resulted in the construction of two pollen percentage diagrams covering the last 8000 years. Additional work was done on the sediments from the two cores TM-I and TM-II that were described according to the characterization system of unconsolidated sediments proposed by TROELS-SMITH (1955) and simplified by AABY in the project guide for IGCP 158 B (BERGLUND 1979)*. Nine sam-

*The sediment description and some additional analyses were performed on Core TM-II by Dr. BENT AABY, Geological Survey of Denmark.

ples from different stratigraphic levels were submitted for 14 C determination thus making it possible to calculate the deposition time (HAKANSSON 1983, 1984) and correlate the sequences.

2 RESULTS and DISCUSSION

The presentation of the time-depth curves lends credence to the idea about the accumulation of different kinds of deposits and the existence of the local vegetation units (Fig.1). Five main stages are delimited with characteristic features and participation of the hydrophyte and hygrophyte pollen types.

Period I - it covers the radiocarbon age interval from 8000 ± 110 B.P. till 4760 ± 80 B.P.

Core TM-I
  445-405 cm/hard grey clay with fine mineral particles and traces from lake gyttja
  405-400 cm/a band of peat
  400-350 cm/calcareous clay-gyttja

Core TM-II
  350-290 cm/As4, Ag++, Ld°
  290-240 cm/As2, Ag1, Ld1, Dh++, Lc+++ with shell remains from freshwater molluscs

The character of the sediments reflects the existence of a lake with a relatively high water level and slow accumulation (DT=45.9 years x cm^{-1}). A small number of samples from Core TM-I proved suitable for pollen analysis with the finding of Typha

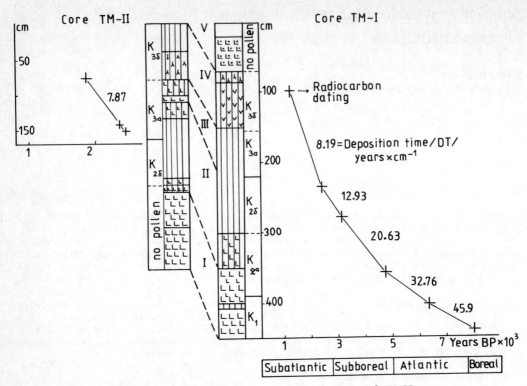

Figure 1. Age/depth curves for cores TM-I and TM-II.

latifolia, T.angustifolia, Sparganium, Myriophyllum, Nuphar with an almost complete lacking of Cyperaceae. The band of peat (Core TM-I) was dated at 6300 ± 65 B.P.

Period II - in terms of radiocarbon years it lasted from 4760 ± 80 B.P. till 2100 B.P. The preservation of the pollen grains was comparatively good.

Core TM-I
     350-300 cm/clay gyttjy and peat with
               mineral particles
     300-210 cm/peat at different degree of
               humification, fragments of
               herbaceous plants with mine-
               ral particles

Core TM-II
     240-230 cm/As3, Thl, Dh+, Ld^o, Lc^o
     230-220 cm/Th3, Asl, Dh+, Lc+, Ag+
     220-135 cm/Th3, Asl, DG+, Ld^o
     135-110 cm/Th4, Dh+, As++

The transition between the first two periods is rather important from stratigraphic point of view. The beginning of the accumulation of Cyperaceae peat could serve as an indication for lowering of the water

level as well as the establishment of optimal conditions for spreading of water and marshy plants. In both diagrams appears the continuous pollen curve of Cyperaceae together with Nymphaea, Nuphar, Potamogeton, Typha latifolia, T.angustifolia, Sparganium, Lythrum and Iris.

The change in the water level after 4760 B.P. had resulted from the natural successions of overgrowing the shores of the lake and the reduced quantities of precipitation feeding the basin. The deposition time was gradually reduced from 20.63 to 8.19-7.87 years x cm^{-1}.

In between 3130 and 2380 B.P. the fen type of vegetation represented by Carex ssp., Filipendula ulmaria, Valeriana officinalis, Potentilla palustris, Parnassia palustris, Menyanthes trifoliata, Polygonum bistorta, Juncus ssp. had been widely distributed. The presence of this vegetation was proved by the finding of rich macrofossil record (STOJANOVA, TONKOV, 1985). Typical phenomenon is the local maximum of Dryopteris thelypteris (2500-2300 B.P.) which had developed near-by the communities of Phragmites and some water macrophytes with floating leaves. The peat layers had become more stable which marked the start of the colonization of the peripheral parts by willows and alders.

Period II - is lasted from 2100 till 1860 $\pm$ 50 B.P.

Core TM-I

    210-150 cm/humified peat with fragments
            of herbaceous plants;
            mineral particles and traces
            from lake gyttja

Core TM-II

    110-105 cm/As4, Dh+, Ld+
    105- 80 cm/Th2, As2, Dh+, Ld+

The deposition of coarse reed and sedge peat was the reason for the poor preservation of the pollen grains in quantity as well. The herbs showed a rise of Poaceae pollen mainly due to the spreading of communities. The decline of all arboreal pollen curves and the deposition of allochthonous material in the peripheral parts presumes deforestation of the surrounding hills by the population of the Thracian and Roman settlements in the vicinity of the marsh.

Period IV - in terms of radiocarbon years it is delimited between 1860 $\pm$ 50 B.P. and 1250 $\pm$ 50 B.P.

Core TM-I

    150- 90 cm/fen-carr peat humified at
            different degree with traces
            from detritus and mineral
            particles
     90- 70 cm/peat mixed with lightbrown
            marl

Core TM-II

     80- 40 cm/Th4, Tl, Dh+, Ag+
     40- 0 cm/Th4, Dh+, Ag++, As+

During this period sedge communities of different species of Carex, Scirpus, Juncus, Heleocharis and shrubs of Salix and Alnus had been distributed, which was confirmed by an increase of the pollen percentage values. Pollen from Alisma, Lythrum, Potamogeton, Hippuris, Epipactis palustris and spores from Sphagnum and Polypodiaceae were also found so that some 1400 years ago in the marsh the main vegetation types had already existed. Since that period pollen of Nymphaea and Nuphar has been lacking in the fossil record but the possibility for their restricted distribution could not be excluded.

The last V period from the development of the marsh started after 1250 $\pm$ 50 B.P. and is represented only in the uppermost

horizone of Core TM-I (hard eroded material without pollen content). The Thickness of this layer varies in the southwestern part from where the material for investigation was taken from an untouched sequence.

This investigation appears as one of the first steps towards complex paleoecological research of reference sites in Bulgaria included in the IGCP Project 158 B

3 ACKNOWLEDGEMENTS

The author is very greatful to Dr. Bent Aaby and Dr. Soren Hakansson as well as to Dr. Elisaveta Bozilova, Prof. Hans-Jürgen Beug and Prof. Björn Berglund for their help and interest in the progress of this research.

4 REFERENCES

BERGLUND,B. (ed.) 1979/1982: Palaeohydrological changes in the temperate zone in the last 15 000 years. Subproject B. Lake and mire environments. Project guide 1-3. LUNDBDS (NBGK - 3009) 626 pp.

HAKANSSON,S., 1983: University of Lund Radiocarbon dates. - In: Radiocarbon, vol. 25, 883-884.

HAKANSSON, S., 1984: University of Lund Radiocarbon dates. - In: Radiocarbon, vol. 26 (3), 400-401.

STOJANOVA,D. & Sp.TONKOV, 1985: Macrofossil plant remains from Tschokljovo marsh-preliminary results. - In: Abstracts of papers and posters, IGCP Project 158 symposium in Bulgaria, Varna 29.IX.-4.X. 1985.

TROELS-MITH,J., 1955: Karakterisering af lose jordater. - Dan. Geol. Und. IV Raekke 3, 1-73.

2 Fluvial environments

Lake, Mire and River Environments, Lang & Schlüchter (eds)
© 1988 Balkema, Rotterdam. ISBN 90 6191 849 9

The evolution of the Thessaloniki-Giannitsa plain in northern Greece during the last 2500 years – From the Alexander the Great era until today

T.A.Astaras & L.Sotiriadis
Department of Geology and Physical Geography, Aristotle University of Thessaloniki, Thessaloniki, Greece

ABSTRACT: The purpose of the study was the establishment of a chronology for the paleo-geographical evolution of the Thessaloniki-Giannitsa plain during the last 2500 years by a combination of information on palynological and radiocarbon data, archaeological and geomorphological/geological evidence and historical records.

The Thessaloniki plain has been formed mainly by the progradation of the deltas of the Axios in the north and the Aliakmon River in the south-west. The positive epeirogenic movement (uplift) which is taking place in this area during the Holocene period is also responsible considerably for the regression of the sea from the plain.

Before the era of the Macedonian king, Alexander the Great (ca. 500 B.C.) the Thermaikos gulf reached Pella and other ancient cities. Later, the continued pro-gradation of the deltas of the Axios and Aliakmon River reduced the size of the gulf and during the period of first Century B.C. and the first Century A.D., the inner part of the gulf was connected to the rest of the Thermaikos gulf by a narrow channel only. During the fifth Century A.D. the inner part of the gulf was cut off completely and formed the Ludias lake (later named Giannitsa lake) which was drained by the Ludias River until the 1930's.

Then, the Giannitsa lake has been completely drained for agricultural purposes. Today, extensive irrigation is used in the cultivation of the plain.

1 INTRODUCTION AND PURPOSE OF THE STUDY

The studies on ancient Macedonian topo-graphy were individual attempts mainly in a single field without reference to other findings and experiences. That is, they were either investigations based on paly-nological and radiocarbon data or on archaeology based on historical records and field evidence.

The objective of the present study was to give an idea for the phases of alluv-iation of the Thessaloniki-Giannitsa plain during the last 2500 years as well as to describe the human intervention in the area during the last 50-60 years.

This goal is achieved by combining paly-nological and radiocarbon data (BOTTEMA 1974) with hydrological-pedological (NEDECO Ltd. et al. 1970) and archaeologi-cal evidence as well as historical records, geomorphological-geological data and information from historical climatology.

2 THE ENVIRONMENT OF THE AREA

The Thessaloniki-Giannitsa (Yannitsa) plain covers about 1700 square kilometers in Macedonia (Fig. 1). It is the largest lowland plain in Northern Greece. It is an alluvial plain occupying an oval depression, eastward of Thessaloniki, sur-rounded on three sides (N, W, S) by semi-cycle hills and mountains (over to 2000 m). To the E-SE the plain meets the Thermaikos gulf (part of North Aegean sea) in a series of deltas. West of the Axios River and to the north, the Paikon Mt. (1080 m) and the Jena Mt. (2180 m, the NE part of the Voras mountainous ridge) appear. Between is the subsidiary depression of the Thessaloniki-Giannitsa plain, the Almopian basin, which drains southward to the Thessaloniki-Giannitsa plain. To the west are the Vermion Mt. (2059 m) and Pieria Mt. (2189 m). The Vermion Mt. is cut off from the Pieria by the Aliakmon gorge. The southern boundary of the plain is formed by a range of low hills stretching eastwards from Pieria Mts. to the Gulf of Thermaikos.

The plain is drained by the large rivers Axios (Vardar) and Aliakmon and smaller Gallikos (Echedorus in Antiquity), Moglenitsa and Ludias.

The Axios and the Aliakmon rivers rise outside the mountains ringing the Thessaloniki-Giannitsa plain and follow complex courses to the sea. The Axios, the largest river, originates in Crna Mts. of Yugoslavia, to the NW of the Voras (Jena) Mt.. The Aliakmon river rises at lake Kastoria to the west of the Vernon-Askion Mts. in NW Greece (Fig. 1).

The Axios river discharge and sediment load is more important than of the Aliakmon. According to THERIANOS (1974, from ALBANAKIS 1985) the maximum discharge in both rivers occurs during March, three months after the maximum precipitation (see below), as a result of the melting of snow in the surrounding mountains. During this peak period the mean monthly discharge of the Axios river exceeds 270 m^3/sec and the discharge of the Aliakmon river reaches 137 m^3/sec. There is a gradual decrease in discharge until August when the minimum is reached (49 m^3/sec for the Axios and 21 m^3/sec for the Aliakmon). According to CONISPOLIATIS (1979, from ALBANAKIS 1985) the annual sediment load for the Axios and Aliakmon rivers are in the order of 300'000 tons and 60'000 tons, respectively, per year.

The Gallikos river, which originates above Kilkis is almost dry in the summer and brings little water during the winter season. It discharges into the Thermaikos gulf. The other two smaller rivers, the Moglenitsa (a tributary of the Aliakmon today) originate in the Almopia and the Ludias in the Giannitsa plain itself.

According to KÖPPEN's classification, the Csa climatic type prevails in the area (Mediterranean climate, humid mesothermal). Only in the surrounding high mountain areas the climate changes to the Cfa type (humid subtropical climate, humid mesothermal, ASTARAS & SILLEOS 1984).

The mean annual rainfall in the town of Giannitsa for the period 1950-1973 (BALAFOUTIS 1977) is 578,63 mm and in the surrounding high mountain areas between 700-1000 mm. The maximum precipitation in the plain appears during the months November to December and accounts for 28-29 % of the annual precipitation. Minimum precipitation is in August.

The prevailing winds, with northerly components, appear during winter and early spring (LIVADAS & SAHSAMANOGLOU 1973) when the rivers have their maximum discharge.

Therefore, the climate of the study area has been influencing the sedimentation in the Thermaikos gulf, either indirectly by freshwater discharge from the surrounding area providing the main source of suspended sediment to the gulf or through the winds, which are the main driving force for currents in the waters of the bay.

Turning to the geology of the area: Mesozoic limestones, marbles schists, phyllites, granites and ultrabasic rocks prevail in the mountains; Neogene to Pleistocene sediments occur in the hills and Holocene alluvial deposits in the plains and valleys (Fig. 2).

The soils of the plain and its neighbourhood are strongly influenced by active erosional and sedimentation processes. The dominant soils in the study area are: redzinas, brown Mediterranean soils in the mountain calcareous areas and podzol acid brown forest soils in phyllite, schists and granite mountain areas. The soils derived from the soft Pleistocene-Neogene deposits are red loam; after man's interference by clearing the vegetation for cultivation, they suffer from down wash and gully erosion, usually during the winter rainfall. The alluvial deposits in the plain and surrounding valley bottoms consist of a variety of soil texture. Usually these soils contain a large percentage of silt and sand.

According to NEDECO (1970) a large part of the soils in the plain is deposited under marine - most probably lagoonal - conditions. Close to the sea, marine sediments are found at/or near the surface. River sediments were brought into the area already during the lagoonal stage. They gradually filled in the lagoon and later the lake, thus giving the Thessaloniki-Giannitsa plain its actual shape. The upper 4 to 10 m consist of freshwater sediments.

The pedogenesis in the plain is young and, consequently, soil formation is limited, and only two soil classes could be distinguished: entisols and inceptisols. The entisols are characterized by the presence of weakly developed subsoils and are sea-bottom, lagoon bottom and lake-bottom soils. The inceptisols were formed by the flood events of the rivers

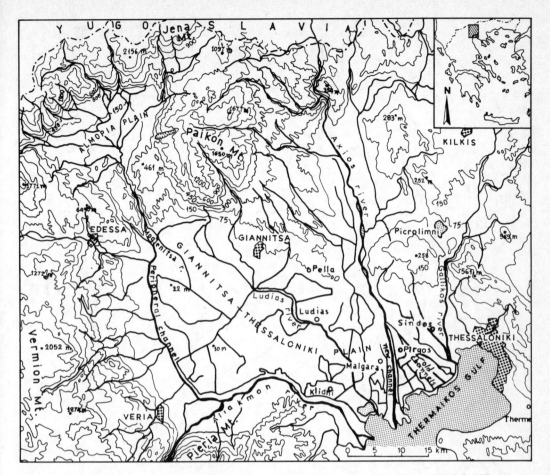

Figure 1. TOPOGRAPHIC MAP OF GIANNITSA - THESSALONIKI PLAIN (NORTHERN GREECE) AND
SURROUNDINGS, REDRAWN FROM THE TOPOGRAPHIC MAP OF THESSALONIKI, GREECE (1:500'000).
EDITED BY THE HELLENIC ARMY GEOGRAPHICAL SERVICE, ATHENS 1979.
(Contour lines in meters; contour interval: 75 m, 150 m or 300 m).

with a high degree of aeration and limited
pedogenesis.

The vegetation of Thessaloniki-Giannitsa
plain is cultivated. Only the saltflats
along the coast show a rather undisturbed
vegetation. Along the sea coast, in the
area which is occasionally flooded by salt
water, halophytic vegetation occurs, such
as Salicornia, Suaeda, Obione etc.
(BOTTEMA 1974).

The cultivated plain is mainly of irrigation agriculture-type with cotton, rice
sugar-beets, tobacco, fruit-trees and
vegetables. Few rain-fed crops (cereals,
almonds, etc.) are concentrated on the

undulating parts of the plain and along
the slopes of the adjacent hill areas.

Along the banks of the streams marsh
plants occur, such as Typha angustifolia,
Phragmites communis and Scirpus lacustris
(BOTTEMA 1974). These plants are also
observed in the drainage ditches in the
center of the plain. They must be the
remnants of the marsh vegetation that
covered a large part of the area at the
beginning of the Century. Apart from the
halophytic vegetations and the marshes the
plain must have been covered with forest
(swamp forest and deciduous trees,
BOTTEMA 1974).

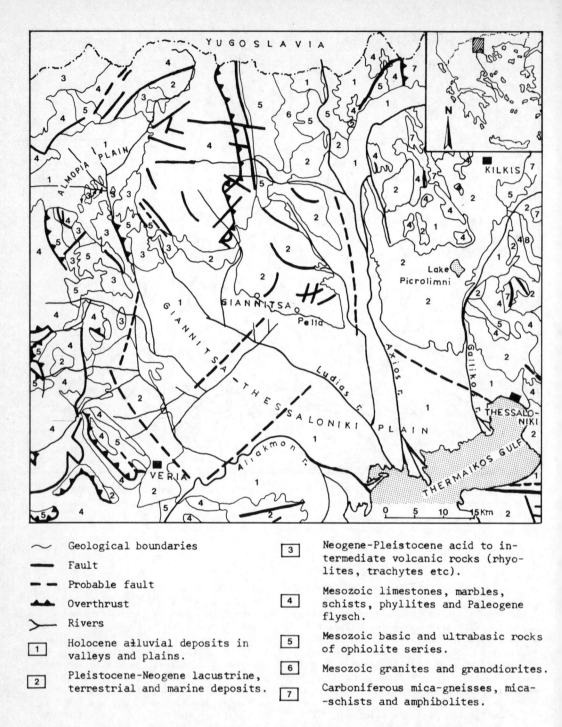

Figure 2. GEOLOGICAL MAP OF GIANNITSA - THESSALONIKI PLAIN (NORTHERN GREECE) and SURROUNDINGS. REDRAWN FROM THE GEOLOGICAL MAP OF GREECE (1:500'000). EDITED BY THE INSTITUTE OF GEOLOGY AND MINERAL EXPLORATION, SECOND EDITION, ATHENS 1983.

Legend:

~ Geological boundaries

— Fault

-- Probable fault

▲▲ Overthrust

>— Rivers

1 Holocene alluvial deposits in valleys and plains.

2 Pleistocene-Neogene lacustrine, terrestrial and marine deposits.

3 Neogene-Pleistocene acid to intermediate volcanic rocks (rhyolites, trachytes etc).

4 Mesozoic limestones, marbles, schists, phyllites and Paleogene flysch.

5 Mesozoic basic and ultrabasic rocks of ophiolite series.

6 Mesozoic granites and granodiorites.

7 Carboniferous mica-gneisses, mica-schists and amphibolites.

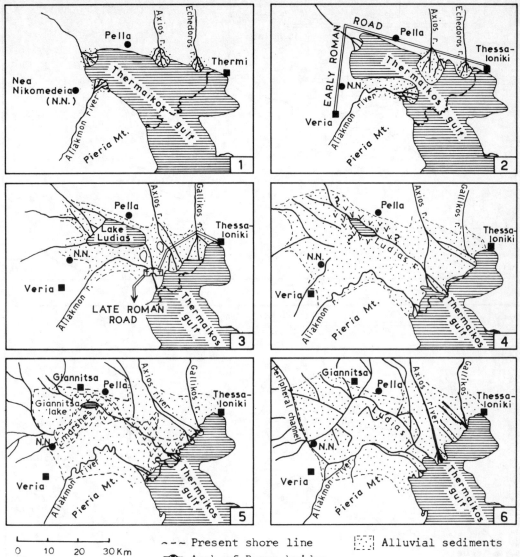

Figure 3. MAP SEQUENCE OF THE PALEOGEOGRAPHY FOR THE PROGRADING ALLUVIATION OF THE
THESSALONIKI - GIANNITSA PLAIN SINCE 500 B.C. (Fig. 3/1 to 3/4 are redrawn from
STRUCK (1908) and BINTLIFF (1975) with modifications). 3/1 - Archaic-Classical period
(circa 500 B.C., before the Alexander the Great era). 3/2 - Early Roman period
(circa 0 B.C./A.D.). 3/3 - Late Roman period (circa 500 A.D.). 3/4 - 1908 A.D.
3/5 - 1926 A.D. 3/6 - Present (1979 A.D.).

The adjacent mountains and hills are
covered by natural vegetation which belongs
(1) to the sub-Mediterranean zone of vegetation with Quercetalia pubescentis
(hilly, submountainous), including both
subzones Ostryo-Carpinion and Quercion confertae; (2) to the mountainous sub-Alpic
zone of vegetation (Fagetalia), with the
dominant characteristic subzone Fagion
moesiaceae (ASTARAS & SILLEOS 1984). The

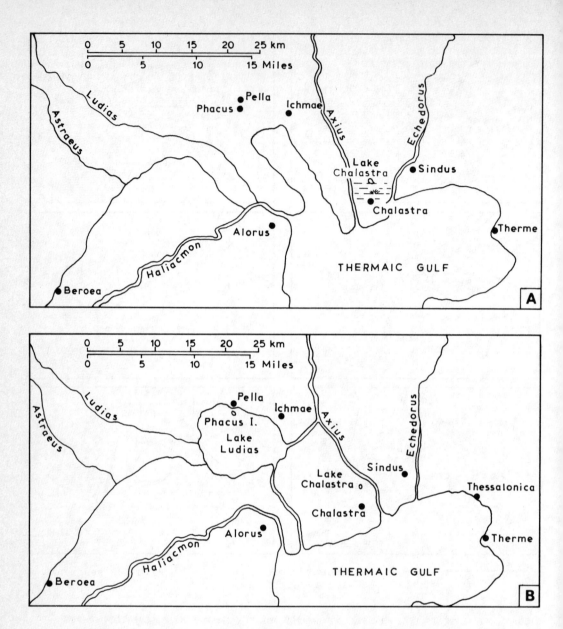

Figure 4. TWO-MAP SEQUENCE FOR THE THERMAIKOS GULF DURING THE PERIOD circa 600-450 B.C. (A) AND THE PERIOD circa 356 B.C. (B). (After HAMMOND & GRIFFITH, 1972).

subzone Fagion moesiaceae covers the highest parts of the Vermion, Pieria, Paikon and Jena mountains. In the lowest parts of the mountains the Quercion confertae alliance appears.

On higher, uncultivated parts of the hills and in the lower mountains the Ostryo-Carpinion is present, with the dominance of Coccifero-Carpinetum and Carpinetum orientalis. Usually the Coccifero-Carpinetum association appears with the characteristic "physiognomy" of Pseudomaqui (Pseudomachie) the dominant and characteristic Mediterranean eco-system.

110

3 PALEOGEOGRAPHIC EVOLUTION OF THE THESSALONIKI-GIANNITSA PLAIN AND THE THERMAIKOS GULF UNTIL THE FOURTH-FIFTH CENTURY B.C.

The Thessaloniki-Giannits plain (Fig. 1) and the greatest part of the Thermaikos gulf is a tectonic depression ("Innen-senke") which has been infilled by sediments and in part by volcanic tuffs along its NW margin (Fig. 2).

The subsidence of this basin, accompanied by marginal and cross faulting has started during the Upper Oligocene-Early Miocene period as a tectonic rift valley ("Graben") and continued until Early Pliocene when it became a closed basin filled with sediments (Fig. 2). During the Quaternary it has remained a tectonic "Graben" (CHRISTODOULOU 1965).

Pliocene and Quaternary are characterized by erosion of the uplands and by producing a topography similar of today. This erosion stripped the uplands of their younger sediment cover and caused incision in more resistant geological units such as the primitive Axios, Aliakmon and other valleys. The accompanying uplift caused valley deepening and the formation of terraces and valley plains that tend to converge down-stream. Within these valleys and along the flanks of the basin tufa terraces were formed locally by depositions from springs.

At the end of the Last Glaciation, the sea-level rose and caused partial flooding of the eroded basin. Along the margins and in the upper reaches of the basin the marine and estuarine deposits are in transition with fluviatile and lacustrine sediments. These conditions continued until at least the beginning of the fourth Century B.C. when Pella, the residence of Alexander the Great, was served by a port (Fig. 1, 2 and 3). According to NEDECO (1970), the latest marine sediments comprise the Cardium edule beds related to a sea-level below the actual one in much of the plain area.

4 PALEOGEOGRAPHIC EVOLUTION OF THE THESSALONIKI-GIANNITSA PLAIN SINCE THE FOURTH-FIFTH CENTURY B.C.

Increasing sedimentary infill of the basin originated mainly from two major rivers: the Axios and the Aliakmon and from Gallikos and Moglenitsa (Fig. 1). This deposition mostly during winter (SHACKLETON 1970) has caused the regression of the sea to the present position. But also the uplift during the Holocene is responsible for the regression. Rejuvenated valleys in Neogene deposits along the northern limits of the Pella area (Fig. 5) support the model of actual uplift. These rejuvenated valleys indicate that the up-lift exceeds the present rise of sea-level by an average of 1 m, or slightly more, per millennium (BINTLIFF 1977).

As long as the sea remained in the basin the rivers Axios and Aliakmon have deposited their sediments directly there (Fig. 3) as fans of delta-like geometry and as sand ridges (spurs) which almost closed-off the course of the Ludias river which was the entrance from the Aegean sea to the gulf. So the brakish lake of Giannitsa remained.

According to NEDECO (1970), which has done some sedimentological work in the plain (200 drillholes were lowered to a maximum depth of 12 m), the first barrier of the lagoon was formed in the western part by the left bank of the Aliakmon river near the present village of Klidhi (Fig. 1). That way, the river could discharge to the sea and by-pass the saline lagoon. It is possible that in the eastern part, near Malgara-Chalastra (Pyrgos)-Sindos, the barrier was formed by the right bank of the Axios-Gallikos system. The two barriers (spurs) formed a line along the villages Klidhi-Malgara-Sindos (Fig. 1). It is plausible that in an early stage a connection did exist between the Aliakmon and the Axios spurs around the village of Klidhi.

Catastrophic events happened in both the Axios and Aliakmon and caused both rivers to flood the lagoonal area and to establish fresh water conditions. And the deposition of typical limnic sediments began. Further growth of the two main deltas led to the formation of a new inner barrier, near the village of Ludias, which, in a later stage, closed the Giannitsa lake. This barrier overlies sediments of the limnic phase. Therefore it must be younger than the outer barrier of Klidhi-Malgara-Sindos.

The increasing input of silt to the plain resulted in the filling-up of the Giannitsa lake and in the establishment of the actual topography (before man-made draining).

With the regression of the sea, the rivers were unable to maintain their courses. The Aliakmon changed its course and often reached the lake of Giannitsa

Figure 5. Rejuvenated valley (downcutting stream course) at the northern limits of the Pella area due to tectonic uplift.

Figure 6. Ruin of a roman bridge (see Fig. 3/3), 3 km east of Klidhi (Fig. 1) and some 6 km inland of the coast. It is one of originally eight to ten arches. It is on the right bank of the present Ludias channel which drains what was, until 1933, the Lake of Giannitsa. According to historians this bridge crossed the outflow of the lake at the time of the Later Roman Empire.

and not the Thermaikos gulf or it joined the Ludias. The Axios has also shifted constantly from east to west and finally discharged partially into the Ludias and into lake Giannitsa. In contrast, the Gallikos did not change its course and discharged to the Thermaikos gulf.

About half a Century ago, the Axios has been brought back by human intervention to its old bed, and the large marshes and the lake Giannitsa have been completely drained for agricultural purposes by replacing the Ludias river by the Ludias drainage channel (Fig. 3/6). Today the Axios and the Ludias discharge together to the Thermaikos gulf. The Axios discharges west of its previous outlet and 3 km east of the Ludias (Fig. 1).

Control dams were built along the Axios and Aliakmon outlet for their control.

5 AGE OF THE PHASES OF ALLUVIATION IN THE THESSALONIKI-GIANNITSA PLAIN SINCE THE FOURTH-FIFTH CENTURY B.C.

The combination of pedological-sedimentological, palynological and radiocarbon data with archaeological evidence from various sites such as Pella, Nea Nicomedeia, Thermi and with records by ancient Greek historians, such as Herodotus, Thucydides, Hecateus, Thephrastus, Aristotle and others (see HAMMOND & GRIFFITH 1972) combined with geomorphological evidence in combination with information from historical climatology (BINTLIFF 1975, 1977; LAMB, 1977) the phases of alluviation of the Thessaloniki-Giannitsa plain are placed in a chronological framework. These data revise and complete STRUCK's (1908) chronology.

At the end of the fifth Century B.C. the Thermaikos gulf extended far into the plain of the Thessaloniki-Giannitsa. So, Thermi, Nea Nicomedeia and Pella are at the sea shore (Fig. 3/1). Pella in the NW corner of the gulf was a seaport at the time of Philip II and Alexander the Great. HAMMOND & GRIFFITH (1979) have produced a different view in their maps (Fig. 4) of the probable shape of the gulf between fifth and fourth Century B.C. Their maps are based mainly on information from the descriptions by the ancient Greek historians (mainly HERODOTUS).

In the first or second Century B.C., between Pella and the lagoon, a marsh was formed (relict pollen conservation in the sediments near Pella (BOTTEMA 1974) support this suggestion). According to HAMMOND & GRIFFITH (1972) an artificial channel of 120 stadia (24 km) still connected the town with open water at this period.

During the period of the first Century B.C. and the first Century A.D. the entrance to the gulf was already narrow (Fig. 3/2).

STRUCK (1908) argues that the Thermaikos gulf had been closed at about 500 A.D. (Fig. 3/3). A shrinking lake, still connected to the sea by the river Ludias, remained. Palynological results by BOTTEMA (1974) and (Fig. 3/4: pollen diagram at Giannitsa) agree with STRUCK's interpretation. Also the change of the Early Roman road connecting Veria (Beroea) with Thessaloniki during the Late Roman period (Fig. 3/2 and 3/3), supports STRUCK's chronology.

The arch of a Roman bridge along the coastal road is visible today 3 km east of Klidhi (Fig. 1 and 6). This bridge crosses the combined outflow of the Aliakmon and Ludias channels in the Late Roman period, probably, and indicates a connection on land between the Axios and Aliakmon deltas which reflects a considerable increase in land surface at that time. This increase of alluviation has taken place rapidly and has possibly been affected by the moister and cooler climate during Late Roman and Medieval periods probably until only a few Centuries ago (BINTLIFF 1975, 1977; LAMB 1977). This wetter climate of the Middle Ages has changed the hydrology and the vegetation by increased erosion and alluviation throughout the valleys and coastal plains. Human influence, by clearing the vegetation for cultivation, may partially be responsible for the surface erosion and consequent alluviation in the plain, but not as much as the climatic causes. Deforestation is greater, more vigorous than ever before at the present time but, even so, stream courses are actively downcutting because their bedloads are less.

After the end of World War I Lake Giannitsa measured about 5 km^2, averaged a 3 m depth and was 4 m above sea-level; but its area was doubled during high water, together with marshes it covered about 90 km^2 (Fig. 3/5).

In 1933, "The Foundation Company, N.Y." drained completely the Giannitsa lake and the surrounding marshes on behalf of the Greek government. First the Axios river was brought back to its old bed and then a "channelization" of the Thessaloniki-Giannitsa plain along the Ludias drainage system (Fig. 1 and 3/6) was achieved.

6 REFERENCES

ALBANAKIS, C., 1985: Monitoring of suspended sediment concentration using optical methods and remote sensing. Unpublished thesis for Ph.D., Dept. of Geography, The University of Nottingham, U.K.

ASTARAS, Th., & N. SILLEOS, 1974: Land classification of part of central Macedonia (Greece). Int. J. Remote Sensing, vol. 5, No. 1, 289-302.

BALAFOUTIS, Ch., 1977: Contribution to the study of climate of Macedonia and Thessaloniki, Doct. thesis, School of Physics and Mathematics, University of Thessaloniki, Thessaloniki.

BINTLIFF, J., 1975: Mediterranean alluvi-
ation: New evidence from Archaelogy.
Proceeding of the Prehistoric Society,
vol. 41, 78-84.
BINTLIFF, J., 1977: Natural environment
and human settlement in Prehistoric
Greece. British Archaeological Reports,
Suppl. Series 28 (1), 35-58 and 68-70.
BOTTEMA, S., 1974: Late Quaternary
vegetation history of northwestern
Greece. PH.D. thesis, pp. 180,
University of Groningen, The Netherlands.
CHRISTODOULOU, G., 1965: Geological struc-
ture of the Thessaloniki-Giannitsa
plain as it follows from the study
of the material provided by three deep
boreholes drilled in the plain. (In
Greek, with English and German summary).
Bull., Geol. Society of Greece, vol. VI,
No. 2, 250-296.
CONISPOLIATIS, N., 1979: Sedimentation
and mineralogy of Thermaikos bay, NW
Aegean sea. Unpublished M.Sc. thesis,
University of Wales.
HAMMOND, N. & G. GRIFFITH, 1972: A history
of Macedonia. Vol. II (550-336 B.C.),
Oxford University Press.
HELLENIC ARMY GEOGRAPHICAL SERVICE
(H.A.G.S.), 1979: Topographic map of
Thessaloniki, Greece, scale 1:500'000,
Athens.
INSTITUTE OF GEOLOGY AND MINERAL
EXPLORATION (I.G.M.E.), 1983: Geologi-
cal map of Greece, scale 1:500'000,
second Edition, Athens.
LAMB, H., 1977: Climate (present, past
and future), Methuen & Co., Ltd., London.
LIVADAS, G. & Ch. SAHSAMANOGLOU 1973:
Wind in Thessaloniki, Greece. Meteoro-
logika, No. 35, Publ. of the Meteorology
and Climatology Department, Aristotle
University of Thessaloniki.
NEDECO, Netherlands Engineering Consult-
ants, The Hague, The Netherlands, 1970:
Regional Development project of the
Salonika (Thessaloniki) plain. Land
Reclamation Service (YEB) of the Ministry
of Agriculture, Athens.
SHACKLETON, N., 1970: Stable isotope study
of the palaeoenvironment of the Neolithic
site of Nea Nikomedeia, Greece. Nature,
227/5261: 943-944.
STRUCK, A., 1908: Makedonische Fahrten.
Zur Kunde der Balkanhalbinsel, Reisen
und Beobachtungen. Herc. Inst. F.
Balkanforschung in Serajewo, pp. 98.
THERIANOS, A., 1974: The annual discharge
and the geographical distribution of the
run-off in the Greek territory. Bull.
Geol. Society of Greece, vol. XI. No. 1,
28-58.

Lake, Mire and River Environments, Lang & Schlüchter (eds)
© *1988 Balkema, Rotterdam. ISBN 90 6191 849 9*

Lake level changes and fluvial activity in the Late Glacial lowland valleys

S.Bohncke & J.Vandenberghe
Institute for Earth Sciences, Free University, Amsterdam, Netherlands

T.A.Wijmstra
Hugo de Vries Laboratory, Section Palynology and Palaeoecology, University of Amsterdam, Amsterdam, Netherlands

ABSTRACT: Phases of alternating dry and wet conditions during the Late Glacial may be caused by geomorphological-hydrological phenomena and by climatic changes. The conditions of humidity in lowland valleys are expressed in lakes of fluvial origin ("valley lakes"), in backswamps on low terraces and in lakes formed in pingo remnants which are also situated in valley depressions. Palaeobotanical, chemical, sedimentological and geomorphological information from the three environments is collected and compared.
A short, but well expressed, dry phase is found in the pingo-remnants as well as on the valley terrace. It represents the Older Dryas. The drought is caused by lowering of the water table due to deep river incisions and thus is of exclusively geomorphological origin. Several fluctuations are recognized during the Alleröd. A second dry phase is generally found at the end of the Alleröd. A third major minimum in the lake levels is reached in the later part of the Younger Dryas and coincides with widespread aeolian dune formation. The latter two dry periods can mainly be attributed to climatic causes.

1 INTRODUCTION

Changes in the palaeo-hydrology of a region have produced most effect in the valleys. The evolution of the water balance is expressed in the palaeo-ecology and river activity as well as in water level fluctuations. Thus palaeo-hydrological conditions may be recorded in organic deposits, the fluvial morphology and the valley lake levels respectively. The combination of these three kinds of information provides the possibility for a detailed paleo-hydrological reconstruction.

Such an approximation has been applied for the evolution of the Dutch and northern Belgian small lowland valleys during the Late Glacial (Fig. 1). The valley geomorphology has been studied in the Mark basin (VANDENBERGHE et al., 1984; VANDENBERGHE AND BOHNCKE, 1985) and the lake levels in three pingo remnants in the basin of the Leuvenumse beek (DE GANS & SOHL, in prep.; BOHNCKE AND WIJMSTRA, in prep.) and on the Drente Plateau (DE GANS & SOHL, 1981), while the palaeo-ecology has been investigated in both areas.

2 PALAEO-MORPHOLOGY AND -ECOLOGY IN THE MARK BASIN

At the end of the Pleniglacial the scarcity of a vegetation cover, as a result of the cold and dry climate, allowed aeolian activity over wide regions. The relatively small river discharges were not able to transport all the sediments which were supplied into the valley especially by wind action. In this way the accumulation by a heavily loaded river of the braided type, resulted in a wide valley plain. On this valley plain abandoned gullies alternated with low undulating dune forms.

The temperature rise at the transition to the Late Glacial coincided with an increase in precipiation. As a reaction to these climatic changes a regional vegetation cover established and aeolian activity nearly stopped. These effects, in turn, provoked drastic changes in the river characteristics and processes. The energy of the river was multiplied, on the one hand, by higher mean discharges and, on the other hand, by the decrease of sediment load due to bank stability and dis-

Figure 1. Location map of the Mark valley (1 = Notsel, 2 = Bergen, 3 = Halle) and the pingo remnants (4 = Uddeler-meer, 5 = Bleeke Meer, 6 = Mekeler-meer).

appearing aeolian supply. As a consequence the river slope was reduced by the establishment of a meandering stream pattern and by vertical erosion. The formation of large meanders at the beginning of the Late Glacial is recognized in many European regions (e.g. GULLENTOPS (1957), DE SMEDT (1973) and DE MOOR & HEYSE (1987) in Belgium; ROSE et al. (1980) in England and KOZARSKI (1983) in Poland). The incision, which started already from the first development of a vegetation cover, induced a contemporaneous lowering of the local water table involving proportional desiccation of the soil and damaging of the vegetation cover. These dry conditions reached a maximum between ca. 11'800 and 11'960 y BP and are reflected by increased aeolian activity and on the terrace, adjacent to the river incision, by slowing down of the peat growth. The latter phenomenon could be observed e.g. at Notsel (Fig. 2) where peat accumulated at a rate of 1 cm in about 90 years in the telmatic and terrestric conditions between 11'600 and 11'960 ± 50 y BP, while in the later limnic conditions between 10'960 and 11'600 ± 50 y BP an aggradation of 1 cm in about 15 years can be inferred (compaction is not considered as it may be supposed equal over the whole section). The renewed

aeolian activity resulted not only in direct filling of the incised valley but also in valley obstruction so that upstream of these dams "valley lakes" were formed in which gyttja has been deposited. The deep but narrow gullies filled quickly after 11'700 y BP and wetter conditions were restored in the areas near the river. At the beginning of the Younger Dryas temporary large floods caused inundations. The later part of the Younger Dryas, however, is characterized by considerable dune formation (VANDENBERGHE et al., in prep.). A Late Glacial incision with an infilling starting in the Older Dryas, Alleröd or Younger Dryas is observed in other regions too (e.g. SEMMEL (1974) in Germany; MUNAUT & PAULISSEN (1973), VANDENBERGHE (1977), VANDENBERGHE & DE SMEDT (1979) and DIRIKEN (1982) in Belgium, DE GANS (1981) and VAN LEEUWAARDEN (1982) in The Netherlands and SZUMANSKI (1973) in Poland).

The palaeo-ecological evolution may be observed at two different sites: in the centre of the valley incision, registrated in the lake gyttjas and peats, and in the valley plains, recorded in the peat on the Pleniglacial terrace.

Registration in the lacustrine sediments starts only from the time of deepest incision, e.g. in the sites Halle (Fig. 2) and Bergen (VANDENBERGHE ET AL., 1984). The local vegetation determines the pollenassemblage to a large extent: from the start of the registration Betula dominates and water plants reach relatively high values. At the end of the Alleröd the local presence of Pinus, proven by the finds of needles, together with a decline in the aquatic species, may indicate an invasion of pine from the river terraces into the valley borders. This spread of Pinus is induced by drier conditions resulting in a lowering of the water table. The first part of the Younger Dryas shows a return to wet conditions. The later part of the Younger Dryas reflects a well-expressed dry phase as can be concluded from the rising curve of Pinus and the decrease of Typha and Potamogeton.

On the Pleniglacial terrace a mire environment existed during the Late Glacial. Since the beginning of the climatic improvement the botanical record shows a progressive evolution from limnic to telmatic and terrestric conditions culminating in the Older Dryas biozone (Fig. 2). This zone is characterized by the absence of aquatic species and the presence of ruderals like Urtica and Solanum. It corresponds with the

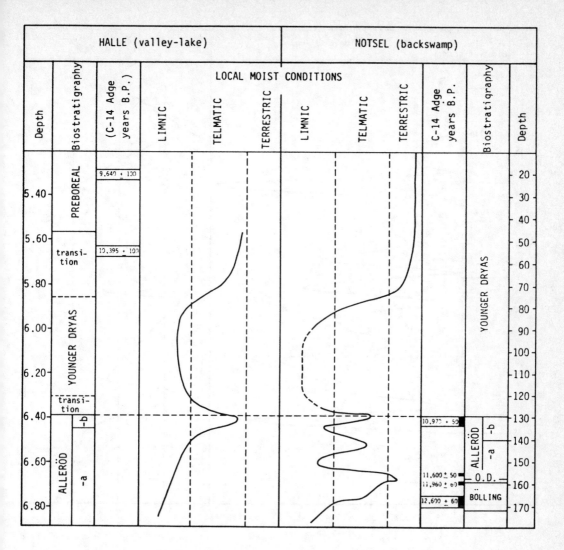

Figure 2. Evolution of the relative humidity in the "valléy lakes" and on the Pleni-glacial terrace in the Mark valley during the Late Glacial.

period of deepest river incision and water table lowering as discussed above. It follows that for the dry conditions in the Older Dryas only geomorphological indications are present. During the Allerød two minor fluctuations of lowered humidity are recorded by a decrease of aquatic species, and the local spread of Hypnaceae peat. At the start of the Younger Dryas river activity on the terrace caused drowning of the peat. Later on, however, fluvial sedimentation was replaced by wet aeolian deposition and finally by dune formation during the later part of the Younger Dryas. This shows

a reestablishment of dry conditions at that time.

3 THE PALAEOHYDROLOGY OF THREE PINGO REMNANTS

Two of the three lakes under study, the Uddelermeer and the Bleekemeer, are located in the basin of the Leuvenumse beek. The Mekelermeer is found on the so-called Drentsche plateau and is also located in a small valley. These lakes are supposed to be remnants of the hydrostatic type (see also MAARLEVELD & VAN DER TOORN (1955), MACKAY (1979) and DE GANS & SOHL (1981)).

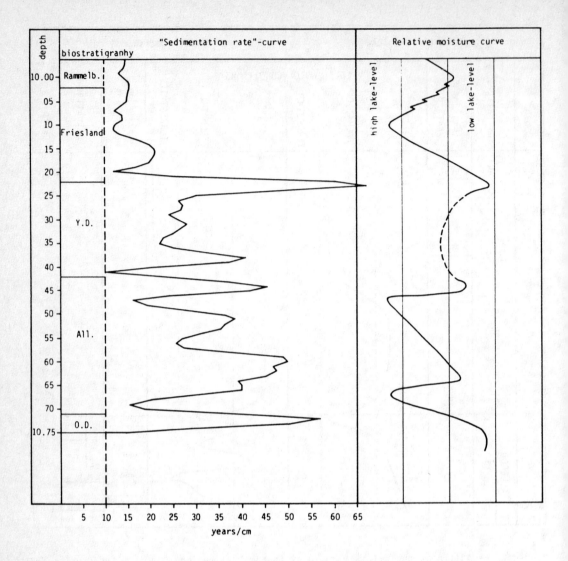

Figure 3. Curves of sedimentation rate and relative moisture in the Bleeke Meer pingo remnant.

The macrobotanical and pollen-analytical data from the three pingo remnants allow the reconstruction of changes in the lake levels. The changes in waterlevel in the three pingo remnants appear to be contemporaneous (BOHNCKE & WIJMSTRA, in prep.). A minimum in the relative moisture curve is reached in the period shortly before the Bölling biozone. Waterlevels restored again during the Bölling but towards its termination lower lake levels returned to the basins (Fig. 3). The water budget in all three lake sequences increased again at the start of the Alleröd. Towards the end of the Alleröd-a phase the Mekelermeer sequence shows evidence for fluctuating water levels, while both the Uddelermeer and the Bleekemeer registered a relative decline in the moisture curve.

Subsequently the second part of the Alleröd (Alleröd-b) shows evidence for a restoration of the water levels. The final stage of the Alleröd, and the transition to the Younger Dryas are marked by a considerable lowering of the lake level recorded in all three lake sequences. The lower part of the Younger Dryas starts with a rise of the water table. Afterwards the

118

waterlevels are lowered again and persisted throughout the Younger Dryas with some minor fluctuations.

The chemical analyses of the lake sediments show an increase in potassium, sodium, magnesium and aluminium during the registered dry periods corresponding to the Older Dryas and the later part of the Younger Dryas biozone. This increase is explained here as the result of an increased aeolian input of unleached skeletal soil into the basins (cf. PENNINGTON, 1977). Also the wet stages in the palaeohydrological record are supported by their chemical characteristics, e.g. an increase in caretenoids (cf. BOHNCKE & WIJMSTRA, in prep.).

Besides a reconstruction of the lake level changes it was hoped to get information on the organic production of the lakes through sedimentation rate curves. The fact that the pollen concentration in a sample is, amongst others, determined by the number of years which are comprised in the sample implies that the pollen concentration holds information about the sedimentation rate. Two complicating factors interfering in the relation AP concentration/sedimentation rate are the anorganic admixture and the increased local pollen precipitation resulting from an inward migration of the lake border shrub vegetation during phases with low lake level. Despite these complications the three sedimentation rate curves, of which one is depicted (cf. Fig. 3), seem to give comparable signals. Eight periods during which a clear decline in the sedimentation rate is registered, e.g. the levels with increasing pollen concentration, can be distinguished. The five most pronounced maxima appear to correlate with periods of low lake level. The conclusion seems acceptable that increased local pollen precipitation, induced by a lowering of the lake level overrules other effects like differnces in accumulation rate (see also Bohncke & Wijmstra, in prep.). This would imply that a "sedimentation-rate" curve is a refined instrument to detect periods of low lake level in Late Glacial lake sequences.

Most of the relatively dry phases as recorded in the pingo remnants find support in the literature, some of them need to be mentioned here. For the Older Dryas period CLEVERINGA et al. (1977) mention a supposed lowering of the lake level in the Uteringsveen pingo remnant (northern Netherlands), and explain it as resulting from an increased evapotranspiration or drainage via the nearby situated brooklet. MENKE (1968) des-

cribes a transition from gyttja into a telmatic sedentation during the Bölling from a Late Glacial lake in northern Germany. KOLSTRUP (1981) suggests drought as the underlying cause for the events registered during the Older Dryas in Denmark. RALSKA-JASIEWICZOWA (1966) finds an expansion of plant communities indicating dry habitats during the younger part of the pre-Alleröd period in the Masurian lake district. The lowering of the lake level at the end of the Alleröd is also registered in southern Sweden (BERGLUND, 1983) and dated at 11'400 - 11'200 y BP. BERGLUND et al. (1984) postulated a lowering of the lake level towards the very end of their late Alleröd-time (11'400 - 11'050 y BP). In Germany, USINGER (1981) explains the recurring hiatus at the end of the Alleröd, and the transition to the Younger Dryas by assuming a decline in the lake level. The Mekelermeer column, taken in the border zone of the lake does not show a fully developed Alleröd pollen record and in comparison with the Uddelermeer and the Bleekemeer diagrams taken in the deepest parts of the lakes, a hiatus at the end of the Alleröd has to be assumed.

Evidence for a restoration of the watertable at the beginning of the Younger Dryas comes from lake Lukcze, Poland (BALAGA, 1982) and is dated to 11'160 - 10'900 y BP. In SW-Sweden BERGLUND et al. (1984) find a sudden climatic deterioration at around 11'050 y BP followed by a rise of the lake level. A distinct lake level lowering towards the end of the Younger Dryas, and the transition to the Preboreal in NW Belgium is found by VERBRUGGEN (1979). This author registers a standstill in the sedimentation of most of the mires under study and some of them even dried out. In SW-Sweden BERGLUND et al. (1984) observe a decline in the water level at about 10'000 y BP. In the Netherlands this transition is quite often badly represented or absent in the pollen record, which can be ascribed to the same mechanism.

4 COMPARISON

A comparison of the periods of low humidity, registered in the pingo remnants, with the dry periods observed in the fluvial system reveals a remarkable synchronism (cf. Fig. 4). The geomorphologically induced dry period, corresponding to the Older Dryas, is also registered in the lake sequences. The relation of these pingo remnants of the

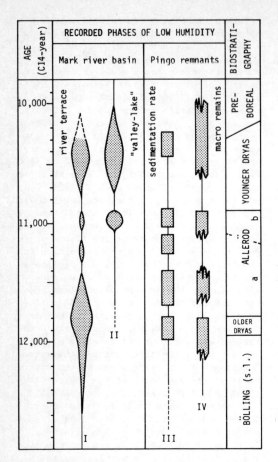

Figure 4. Comparison of the periods of recorded low humidity in the Mark valley and the lakes of the three pingo remnants.

hydrostatic type to adjacent fluvial systems explains the contemporaneous decline in the watertable. The climatically determined dry period at the end of the Alleröd is expected to have a regional effect and is recorded in all three environments (backswamp, valley lakes and pingo remnants). The increased wet conditions in the lower part of the Younger Dryas are also registered in all three environments. This increase in wet conditions is not necessarily climatically induced but can be explained by an increase in the watertable due to quick aggradation of river sediments and higher peak discharges. The transition from rather wet conditions in the lower part of the Younger Dryas towards dry conditions during the later part of the Younger Dryas is again observed in all three environments.

Only within the deepest parts of the pingo remnants and valley-lakes pollen registration continued during this period while the borders and river terraces emerged above the water table. For this part of the Younger Dryas a drier climate is assumed, which led to inland and river dune formation

ACKNOWLEDGEMENTS

Mr. H. Sion and Mr. C. van der Bliek kindly prepared the figures. Mrs. Romée de Vries typed the manuscript.

REFERENCES

BALAGA, K. 1982: Vegetational history of the lake Lukcze environment (Lublin Polesie, E. Poland) during the Late-Glacial and Holocene. Acta Palaeobotanica XXII: 5-22.

BERGLUND, B.E. 1983: Palaeoclimatic changes in Scandinavia and on Greenland. A tentative correlation based on lake and bog stratigraphical studies. Quaternary studies in Poland 4: 27-44.

BERGLUND, B.E., LEMDAHL, G., LIEDBERG-JÖNSSEN, B., and PERSSON, T., 1984: Biotic response to climatic changes during the time span 13'000 - 10'000 B.P. A case study from S.W. Sweden. In: N. Mörner and W. Karlén (eds.). Climatic changes on a yearly to millenial basis, Reidel, Dordrecht: 25-36.

DE GANS, W. 1981: The Drentsche Aa valley system. Doctor's thesis, Free University Amsterdam: 132 pp.

DE GANS, W. % SOHL, H. 1981: Weichselian pingo remnants and permafrost on the Drente plateau (The Netherlands). Geologie en Mijnbouw 60: 447-452.

DE MOOR, G. & HEYSE, I. 1978: De morfologische evolutie van de Vlaamse Vallei. De Aardrijkskunde: 343-375.

DE SMEDT, P. 1973: Paleogeografie en kwartairgeologie van het confluentiegebied Dijle-Demer. Acta Geographica Loveniensia 11: 141 pp.

DIRIKEN, P. 1982: Postglaciale evolutie van de Molenbeek-Mombeekvallei (Belgisch Haspengouw). Natuurhistorisch Maandblad 71: 8-19.

GULLENTOPS, F. 1957: Quelques phénomènes géomorphologiques depuis le Pléni-Wurm. Bulletin Société belge de Géologie 66: 86-95.

KOLSTRUP, E. 1981: Late-Glacial pollen diagrams from Hjelm and Draved Mose (Denmark)
with a suggestion of the possibility of
drought during the Earlier Dryas. Rev. of
Palaeobot. Palynol. 36: 35-63.

KOZARSKI, S. 1983: River channel changes in
the middle reach of the Warta valley,
Great Poland lowland. Quaternary Studies
in Poland 4: 159-169.

MAARLEVELD, G. & VAN DEN TOORN, J. 1955:
Pseudo-sélle in Noord Nederland. Tijdsch.
K.N.A.G. (2e reeks) 72: 342-360.

MACKAY, J.R. 1979: Pingos of the Tuktoyaktuk
peninsula area, Northwest territories.
Géographie Physique et Quaternaire
XXXIII(1): 3-61.

MENKE, B. 1968: Das Spätglazial von Glüsing.
Eiszeitalter und Gegenwart. 19: 73-84.

MUNAUT, A. & PAULISSEN, E. 1973: Evolution
et paléo-écologie de la vallée de la
Petite Nèthe au cours due Post-Würm (Belgique). Annales de la Société Géologique
de Belgique 96: 301-348.

PENNINGTON, W. 1977: Lake sediments and the
Late Glacial environment in northern
Scotland. In: J. Gray & J. Lowe, Studies
in the Scottish Late Glacial environment,
Pergamon Press, Oxford: 119-141.

RALSKA-JASIEWICZOWA, M. 1966: Bottom-sedi-
ments of the Mikolajki Lake (Masurian
Lake District) in the light of palaeobotanical investigations. Acta Palaeobotanica VII-2: 1-118.

ROSE, J., TURNER, C., COOPE, G.R. & BRYAN,
M.D. 1980: Channel chantes in a lowland
river catchment over the last 13'000 years.
- In: R. Cullingford, D. Davidson & J.
Lewin (eds.), Timescales in Geomorphology,
Wiley, New York: 159-176.

SEMMEL, A. 1972: Untersuchungen zur jung-
pleistozänen Talentwicklung in deutschen
Mittelgebirgen. Zeitschrift für Geomorpho-
logie N.F., Suppl. Band 14: 105-112.

SZUMANSKI, A. 1983: Paleochannels of large
meanders in the river valleys of the Polish Lowland. Quaternary studies in Poland
4: 207-216.

USINGER, H. 1981: Ein weit verbreiteter Hiatus in spätglazialen Seesedimenten: mög-
liche Ursache für Fehlinterpretation von
Pollendiagrammen und Hinweis auf klima-
tische verursachte Seespiegelbewegungen.
Eiszeitalter und Gegenwart. 31: 91-107.

VANDENBERGHE, J. 1977: Geomorfologie van de
Zuiderkempen - Verhandelingen Koninklijke
Academie voor Wetenschappen, Letteren en
Schone Kunsten van Belgie, Klasse der
Wetenschappen 140-166.

VANDENBERGHE, J., VANDENBERGHE, N. &
GULLENTOPS, F. 1974: Late Pleistocene
and Holocene stratigraphie in the neighbourhood of Brugge. Mededelingen Koninklijke Academie voor Wetenschappen,
Letteren en Schone Kunsten van Belgie,
Klasse der Wetenschappen XXXVI-3: 77 pp.

VANDENBERGE, J. & DE SMEDT, P. 1979:
Palaeomorphology in the eastern Scheldt
basin (Central Belgium). Catena 6: 73-105.

VANDENBERGHE, J., BEYENS, L., PARIS, P.,
KASSE, C., and GOUMAN, M. 1984: Paleo-
morphological and -botanical evolution
of small lowland valleys (a case study
of the Mark valley in northern Belgium).
Catena 11: 229-238.

VANDENBERGHE, J. & BOHNCKE, S. 1985: The
Weichselian Late Glacial in a small lowland valley (Mark river, Belgium and
The Netherlands). Bulletin de l'Association Française d'Etudes Quaternaires, in
press.

VAN LEEUWAARDEN, W. 1982: Palynological and
macropalaeobotanical studies in the development of the vegetation mosaic in
Eastern Noord-Brabant (The Netherlands)
during Late Glacial and Early Holocene
times. Doctor's thesis, Rijks Universiteit Utrecht: 203 pp.

VERBRUGGEN, C.L.H. 1979: Vegetational and
Palaeoecological history of the Late Gla-
cial period in Sandy Flanders (Belgium).
In: Y. Vasari, M. Saarnisto and M. Seppälä
(eds.), Palaeohydrology of the Temperate
Zone, Acta Universitatis Oulensis, A-82
Geol. 3: 133-142.

Lake, Mire and River Environments, Lang & Schlüchter (eds)
© 1988 Balkema, Rotterdam. ISBN 90 6191 849 9

Outline of river adjustments in small river basins in Belgium and the Netherlands since the Upper Pleniglacial

P.Cleveringa
Geological Survey of the Netherlands, Haarlem, Netherlands

W.De Gans
Geological Survey of the Netherlands, District West, Alkmaar, Netherlands

W.Huybrechts
Geografisch Instituut, Vrije Universiteit Brussel, Brussels, Belgium

C.Verbruggen
Geologisch Instituut, Rijks Universiteit Gent, Gent, Belgium

ABSTRACT: Small river-basins in lowland areas of Belgium and The Netherlands show four episodic river adjustments since the Upper Pleniglacial, dated at the end of the Pleniglacial/early Late Glacial and in the Late Dryas, Boreal, and Subboreal/Subatlantic. These adjustments are discussed and explained as water discharge fluctuations due to changes in precipitation and/or evapotranspiration. In relation to the river adjustments, tentative curves of precipitation, evapotranspiration, and water discharge over the last 15'000 years are presented.

INTRODUCTION

In past decades a number of papers (e.g. FISK 1944; ANDEL 1950; LEOPOLD & WOLMAN 1957. 1960; DOEGLAS 1962) and handbooks (LEOPOLD 1964; SCHUMM 1977; GREGORY 1977) dealing with the sedimentology, hydrology, and fluvial regimes of river-basins have been published. Recent research has been concerned with a reconstruction of the hydrologic regimes of past fluvial environments, because an interrelationship with climatic fluctuations and human impact has been assumed (DE GANS 1981; BURRIN & SCAIFE 1983; GREGORY 1983; KOZARSKI 1983; VANDENBERGHE 1984; HUYBRECHTS 1985; JAGERMAN et al. in prep.). Meanwhile, attempts have been made to establish a general scheme against the background of which the changes in river activity can be understood (KNOX 1975) and correlated (STARKEL 1983). The present paper deals with the general character of the fluvial activity in small river-basins during the Upper Pleniglacial, Late Glacial, and Holocene for a limited area (Belgium and The Netherlands) with a low relief and more or less homogeneous climatic conditions both in the past and at present. The drainage basins are all situated in unconsolidated rocks of Pleistocene and/or Tertiary age.

SYNOPSIS OF RESEARCH IN SMALL LOWLAND RIVER BASINS

The authors' investigations in the Drentsche Aa, the Leuvenumse Beek, and the Mark valleys (Fig. 1) have demonstrated that responses of river systems to changing environmental conditions are easily recognizable in relatively small drainage basins. A longitudinal profile and a cross-section of the Leuvenumse Beek (Figs. 2 and 3) have for instance revealed several sequences of cutting and filling. Because of the occurrence of organic beds in these sequences, pollen analysis and radiocarbon dating could be used to obtain a better understanding of the chronology of fluvial activity in small lowand rivers. However, the available data confirm our supposition that in practice it is very difficult to judge whether one is dealing with (episodic) incissions or with alternating phases of erosion and aggradation. Therefore, following GREGORY (1977), KOZARSKI (1977, 1983), and STARKEL (1983) we have adopted the term river adjustment for the above-mentioned dynamic processes in fluvial systems.

Additional data provided by well-documented research on fluvial activity in small lowland rivers in Belgium and The Netherlands have revealed that river

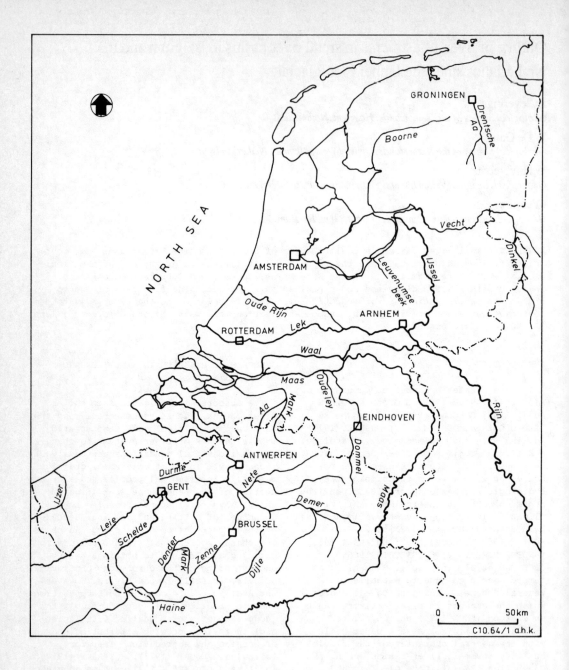

Figure 1. Location of the investigated lowland river basins in Belgium and The Netherlands.

adjustments occurred episodically during the last 15'000 years (Fig. 4). The recognized river adjustments during the last 15'000 years are associated with the following periods:

Periods of river adjustment	Number of basins with river adjustment	
1. Subboreal/Subatlantic	7	
2. Boreal	7	
3. Late Dryas	5	
4. Allerød	7)	
5. end of Pleniglacial/ early Late Glacial	) 6)	13

It should be kept in mind that river adjustments are dated in two ways yielding two groups: those for which a true age is 'known', and those for which only a terminus ante quem date is available.

Because the Allerød adjustment has been established in all basins as a terminus ante quem dating and in one specific basin has not been found in connection with the end Pleniglacial/early Late Glacial adjustment, it was assigned to the latter. This period of river adjustment appears to have been recorded the most often since it was recognized by most of the authors; it was found in 13 of 16 river-basins. The other periods of river adjustments have been encountered less frequently; the Late Dryas adjustment in five, and the Boreal in seven basins. The Subboreal/Subatlantic adjustments has been found in seven out of the 16 basins, and unlike the preceding adjustments has often been registered as a series of adjustments over a long period of time. Because the river adjustments are the result of a continuous adaptation of the fluvial system to changes in water and sediment discharges (LEOPOLD & MADDOCK 1953; SCHUMM 1960), we assume that they represent adjustments to changing climate and vegetation (= environmental) conditions. Before discussing the causal relationships between changes in environmental conditions and river adjustments, it will be useful to give a description of development in vegetation cover and climate during the last 15'000 years.

EVOLUTION OF VEGETATION AND CLIMATE DURING THE LAST 15'000 YEARS

Palaeobotanical research - especially pollen analysis - has yielded sufficient information about the history of the post-Pleniglacial vegetation to permit the reconstruction of vegetation pattern (rate of covering by herbs and/or trees) and temperature changes (expressed as average july $^\circ$C).

From the absence of organic material and on the basis of lithological information it is supposed that the period between 15'000 and 13'000 BP was dry and that no closed vegetation cover was present (KOLSTRUP, 1980). After 13'000 BP (none of the oldest ^{14}C dating is earlier than 12'500 BP) Artemisia, Salix sp., and Betula nana increased, which suggests a rise in temperature. The presence of Juniperus and Hippophae indicates the advance of the tree line, because both species are characteristic of areas immediately adjacent to the polar forest limit (VAN DER HAMMEN 1951). The presence of a number of light-demanding species shows that the forest cover was quite open.

During the Allerød the successive rise of the Betula and Pinus pollen curves and the decline of the herb pollen values reflect continuing increase in the rate of the vegetation cover.

However, we agree with COOPE (1970) that the temperature during the Late Glacial reached an optimal level rather early. i.e., prior to the Allerød. Therefore, in agreement with COOPE (1975), VERBRUGGEN (1979) and KOLSTRUP (1980), we assume that real climatic regression did not occur in the Early Dryas.

A decline in the Pinus pollen curve and a rise of the herb pollen values indicate that the vegetation became more open during the Late Dryas. Geomorphologic data (e.g. the presence of ice-crack casts and solifluction deposits) and vegetation development indicate a temperature drop (VAN DER HAMMEN 1951; VAN DER HAMMEN and WIJMSTRA 1971; CASPARIE and TER WEE 1981) for that period.

At the beginning of the Holocene the temperature rose again which is reflected in an increase of pine and the gradual decline of birch. The occurrence of

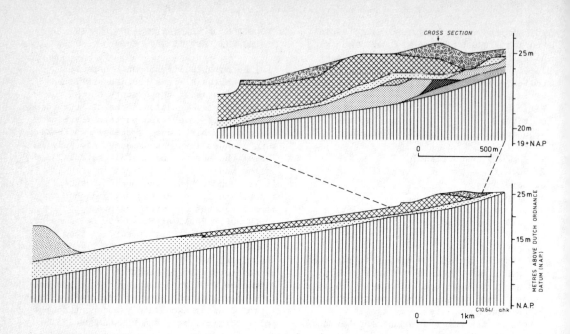

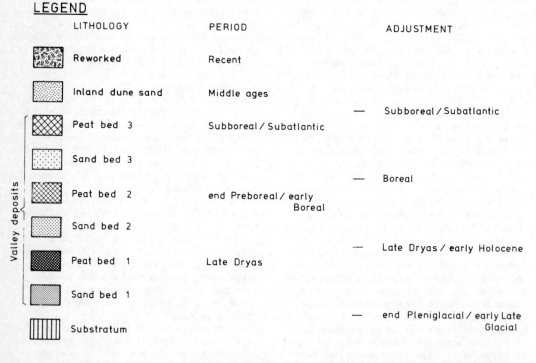

LEGEND

	LITHOLOGY	PERIOD	ADJUSTMENT

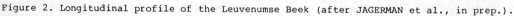

Reworked — Recent

Inland dune sand — Middle ages

— Subboreal/Subatlantic

Peat bed 3 — Subboreal/Subatlantic

Sand bed 3

— Boreal

Peat bed 2 — end Preboreal/early Boreal

Sand bed 2

— Late Dryas/early Holocene

Peat bed 1 — Late Dryas

Sand bed 1

— end Pleniglacial/early Late Glacial

Substratum

Valley deposits

Figure 2. Longitudinal profile of the Leuvenumse Beek (after JAGERMAN et al., in prep.).

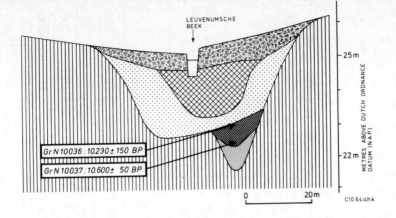

Figure 3. Cross-section of the Leuvenumse
Beek valley (after JAGERMAN et al., in prep.). See Fig. 2 for legend.

Corylus, Quercus, and Ulmus and the decline of the Pinus pollen values during the Boreal indicate the development of deciduous forest at the cost of pine woods. The presence of Viscum and Hedera in the Atlantic period are indicative of a further rise of the temperature. At the transition of the Atlantic to the Subboreal, the first signs of human activity become visible in the pollen diagrams. The gradual rise of such herbs as Artemisia, Rumex, Plantago, Ericaceae, and Gramineae is indicative. Detailed investigations on peat growth (CASPARIE 1972; VAN GEEL 1978; DUPONT 1985) indicate that there was a slight drop of the temperature during the Subboreal/Subatlantic. The rates of vegetation development and temperature changes are shown in relative curves in Fig. 5.

RIVER ADJUSTMENTS AND CLIMATIC PARAMETERS

The episodic character of the river adjustments must be seen in relation to the evolution of climate (temperature) and vegetation during the last 15'000 years. Because this analysis is restricted to small rivers, long-term and long-distance factors are not considered. The causes of river adjustments must be restricted to those parameters controlling the discharge. The discharge itself is influenced by a number of variables such as the intensity and duration of precipitation, the type of vegetation and litter-layer (interception), the infiltration capacity of soils, and the density of the substrate (LOCKWOOD 1983; ALLEN 1977). However, the total amount of water that becomes available for river discharge is strongly dependent on the amount of precipitation and the degree of evapotranspiration (KNOX 1972; ROSE et al. 1980; GREGORY 1977, 1983).

The increased water discharge at the turn of the Pleniglacial/early Late Glacial adjustment is consistent with an increase in temperature. An improvement of the climate and the presence of a non-adapted vegetation due to some delay in tree migration, influence both run off (sediment yield) and infiltration capacity. Delayed migration also leeds to a low evapotranspiration rate. However, a retardation of vegetation development combined with a rise of the temperature (COOPE 1970, 1975) cannot explain the increase in water discharge. In our opinion increased precipitation and the absence of a more or less closed vegetation cover were responsible for the adjustment. Further grounds for the assumption of a considerable increase in precipitation include: 1) the knowledge that the last phase of the Pleniglacial was dry; 2) the widespread presence of organic deposits in shallow depressions of early Late Glacial and Allerød age, indicating a high ground-water level immediately after the Pleni-

127

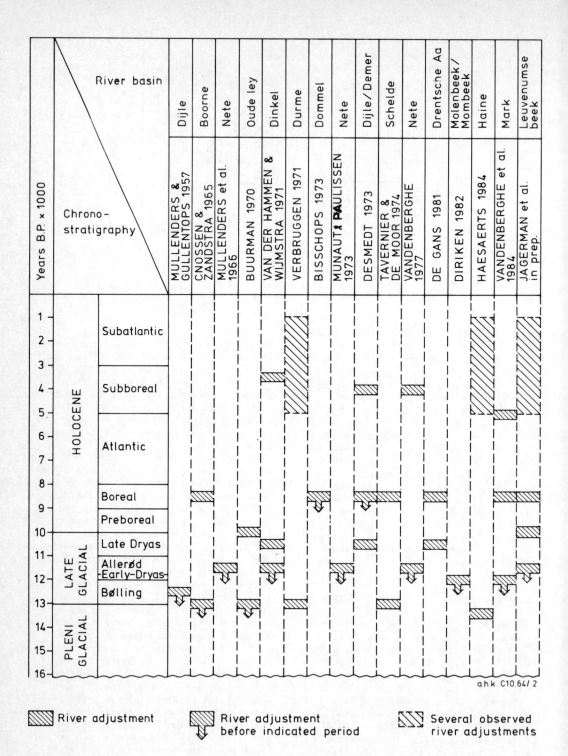

Figure 4. Registered river adjustments in lowland river basins in Belgium and The Netherlands.

128

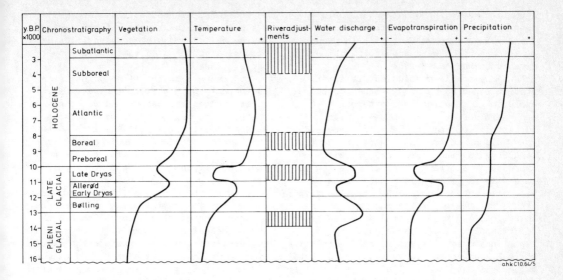

Figure 5. Tentative relative curves of climatic parameters related to river adjustments since the Upper Pleniglacial.

glacial; and 3) the early onset (14'000 y BP) of deglaciation in Scandinavia and the Alps, which could have resulted in a considerable increase of atlantic conditions in the low countries since the Late Glacial. The Late Dryas adjustment was caused by a decrease of evapotranspiration due to a substantial drop of the temperature, followed by vegetation regression. The extent to which the presence of permafrost influenced both adjustments is still an open question. For the Boreal adjustment, a direct relationship with an increased discharge does not seem to be indicated by the climatic parameters. However, one ecological phenomena during the Boreal, namely the change in vegetation from pine forest to deciduous forest has been thought to lead to reduced evapotranspiration (LOCKWOOD 1983).

It is conceivable that this slightly enhanced water discharge could, in combination with a low sediment yield, ascribable to the dense plant cover and to soil-forming processes, lead to river adjustment. Nevertheless, the Boreal adjustment should perhaps be explained as a "base level" adjustment. Although we agree with BLOOM (1969) that with respect to sea-level changes the base-level-concept represents a hypothetical point in a fluvial

system, a point "which never could be reached", it seems, however, possible that in some of the investigated drainage basins erosion continued until infilling was induced by rising sea-level.

At the onset of the Subboreal/Subatlantic river adjustment there is only a slow increase of water discharge associated with a slight temperature drop combined with a slightly enhanced precipitation. At the same time there was an increase of anthropogenic activity. Although the effects of deforestation must have been restricted to small local areas, disturbances of the water circulation and microclimate as well as the degradation of soils must have influenced fluvial activity and evapotranspiration. However, these effects too must have been local and limited. On the whole, both the impact of human activity and the changes in climate led to increased water discharge, but these changes were so gradual and the character of the impact so local that the threshold for the induction of a manifest adjustment was never crossed in many basins and certainly not in all at the same time. The direct effect of clearing on the discharge (run-off) must have been much more pronounced, especially during the second half of the Subatlantic. Up to now it has only been possible to

129

establish these Subboreal/Subatlantic adjustments by detailed investigation of the entire basin (JAGERMAN et al., in prep.) or in connection with archaeological excavations (HAESAERTS 1984).

As a reasonable conclusion, it is stated, that in our area of study, based on the evolution of vegetation and temperature and the observation on episodic river adjustment two main causes for increased water discharge are discerned for the last 15'000 years: increased precipitation and decreased evapotranspiration. This led us to add in Fig. 5, besides the curves of vegetation and temperature some tentative curves for the precipitation, evapotranspiration, and water discharge. Because the relationship between these curves is of importance only, use was made of a relative scale i.e., without indication of absolute values. We have decided on a time limit for these curves at around 2000 y BP (Roman Period) to restrict the discussion to the time before major human influence occurred.

CONCLUSIONS

In the time period since 15'000 y BP four river adjustments have been documented. The Leuvenumse Beek valley is the only basin in which a complete series of all four has been registered (JAGERMAN et al., in prep.). For the other 15 basins under discussion (Fig. 4) the records are incomplete. In the Dinkel valley (VAN DER HAMMEN et al. 1971), the Dijle valley (DE SMEDT 1973) and the Mark valley (VANDENBERGHE et al. 1984) three adjustments out of four have been recognized. In the other basins less river adjustments have been described.

The differences between the river basins under study can be explained in several ways. - In the first place, there are good reasons on which to assume that not all of the river basins reacted with the same intensity, for the same length of time and to the same degree to changing (climatic and/or vegetational) conditions. - In the second place, not all investigators find all river adjustments, the number detected is depending on the methods used and on the scientific objective of the investigation. The different approaches for the respective studies (involving different parts of the basins and the use of specific methods) are of an important role. - Also, the stratigraphic sequences of drainage basins are incomplete because of local dominance of erosional discontinuities. And sedimentation in the lower parts of the basins may be associated with erosional features occurring at the same time in the upper parts.

With respect to the first point it may be added that the reaction of rivers to environmental changes depends strongly on whether threshold conditions are crossed (SCHUMM 1973, 1977; KOZARSKI 1983; STARKEL 1983). Whether and when this occurs in a basin is closely related to the prevailing local conditions, such as bedrock characteristics, morphology, slope, river gradient. In view of the many possible variations in these 'local' conditions, a great diversity of reactions of rivers to the changing environmental conditions is to be expected. However, this does not exclude the existence of a major force governing the adjustments.

A more detailed and complete explanation of the changing fluvial dynamics in a larger area will require the systematic collection of more and well-dated field evidence in different 'fluvial environments'. Such data can be expected to provide a better understanding of the discrepancies and time-shifts that seem to occur between river systems in the same but also in different climatic regions and between river systems differing in size, and might make it possible to distinguish the effects of short-term climatic events (flooding) from major secular episodes of climatic change (shifts in circulation). Finally, in our opinion river adjustments are recognized more readily in rivers with a restricted drainage basin than in larger streams and this holds especially true for the effects of short-term fluctuations of the climate, changes in vegetation, and land use.

ACKNOWLEDGEMENTS

We are indepted to Dr. D. BEETS and Ing. J. DE JONG for their critical reading of the manuscript and suggestions for improvement, to Mr. A KOERS, who prepared the graphs and to Mrs. JOUINI-SPRUIJT, Miss VREUGDEN-BURG and Mr. KRAAKMAN for typing many versions of the manuscript. Mrs. I. SEEGERS read the English text.

REFERENCES

ALLEN, J.R.L. 1977. Changeable rivers: some aspects of their mechanic and sedimentation. In (K.J. GREGORY ed.): River channel changes: 15-46.

ANDEL, Tj.H. van 1950. Provenance, transport and deposition of Rhine sediments. Thesis. University of Groningen, 129 pp.

BISSCHOPS, J.H. 1973. Toelichting bij de geologische kaart van Nederland 1:50'000. Blad Eindhoven Oost (51 O). Rijks Geologische Dienst, Haarlem, 132 pp.

BLOOM, A.L. 1969. The surface of the Earth. Prentice-Hall, Inc., New Jersey, 151 pp.

BURRIN, P.J. & R.G. SCAIFE 1983. Aspects of Holocene valley sedimentation and floodplain development in southern England. Proc. Geol. Ass., 95 (1): 81-96.

BUURMAN, P. 1970. Pollen analysis of the Helvoirt River valley. Geologie en Mijnbouw, 49 (5): 381-390.

CASPARIE, W.A. 1972. Bog development in Southern Drenthe (The Netherlands). Thesis, University of Groningen, 271 pp.

CASPARIE, W.A. & M.W. TER WEE, 1981. Een-Schipsloot. The geological-palynological investigations of a Tjonger Site. Palaeohistoria 23: 29-44.

CNOSSEN, J. & J.G. ZANDSTRA, 1965. De oudste Boorneloop in Friesland en veen uit de Paudorftijd nabij Heereveen. Boor en Spade, XIV: 62-87.

COOPE, G. 1970. Climatic interpretations of Late Weichselian Coleoptera from the British Isles - Revue Geogr. Phys. Geol. Dyn. 12: 149-155.

COOPE, G.R. 1975. Climatic fluctuations in north-west Europe since the last interglacial indicated by fossil assemblages of Coleoptera. In: (A.E. WRIGHT & F. MOSELEY, eds.). Ice Ages, ancient and modern: 153-168.

DE SMEDT, P. 1973. Paleogeografie en kwartairgeologie van het confluentiegebied Dijle-Demer. Acta Geogr. Lovan. Vol. 11, 141 pp.

DIRIKEN, P. 1982. Postglaciale paleo-ecologische evolutie van de Molenbeek-Mombeek vallei (Belgisch Harpengouw). Natuurhist. Maandblad 71 (1): 8-19.

DOEGLAS, D.J. 1962. The structure of sedimentary deposits of braided rivers. Sedimentology, 1: 167-190.

DUPONT, L.M. 1985. Temperature and rainfall variation in a raised bog ecosystem. Thesis. University of Amsterdam, 62 pp.

FISK, H.N. 1944. Geological investigation of the alluvial valley of the lower Mississippi River. Mississippi River Commission, Vickbury, Miss., 82 pp.

GANS, W. de 1981. De Drentsche A. valley system. Thesis. Free University, Amsterdam, 132 pp.

GEEL, B. van 1978. A palaeoecological study of Holocene peat bog sections in Germany and The Netherlands. Rev. Palaeobot. Palynol., 25: 1-120.

GREGORY, K.J. & D.E. WALLING, 1973. Drainage basin form and pocess. A geomorphological approach. Edw. Arnold, London, 458 pp.

GREGORY, K.J. 1977. River Channel Changes. Wiley, Chichester, 448 pp.

GREGORY, K.J. (ed.) 1983. Background to palaeohydrology: A perspective. Academic Press, London, 237 pp.

HAESAERTS, P. & J. de HEINZELIN, 1979. Le site palaeolithique de Maisières-Canal. Dissertationes Archaeologicae Gandenses, 19, Brugge.

HAESAERTS, P. 1984. Les Formations fluviatiles pleistocènes du bassin de la Haine (Belgique). Bull. de l'AFEQ, 19-26.

HAMMEN, T. VAN DER, 1951. Late Glacial flora and periglacial phenomena in The Netherlands. Leidse Geol. Mededelingen, XVII: 71-183.

HAMMEN, T. VAN DER & T.A. WIJMSTRA, 1971. The Upper Quaternary of the Dinkel valley. Med. Rijks Geologische Dienst, N.S. 22. 213 pp.

HEINZELIN, J. de, P. HAESAERTS & J. DELAET 1977. Le Gue du Plantin. Dissertationes Archaeologicae Gandenses, XVIII, 58 pp.

HUYBRECHTS, W. 1985. Morfologische evolutie van de riviervlakte van de Mark (Geraardsbergen) tijdens de laatste 20'000 jaar. Thesis. Brussel, 250 pp.

JAGERMAN, M.P., P. CLEVERINGA & W. DE GANS (in prep.). Aspects of Late Glacial and Holocene development of the Leuvenumse Beek valley (The Netherlands).

JONG, J. de 1981. Chronostratigraphic subdivision of the Holocene in The Netherlands. In: (J. MANGERUD, H.J.B. BIRKS & K.D. JAEGER, eds.) Chronostratigraphic subdivision of the Holocene. Striae, vol. 1: 71-79.

KNOX, J.C. 1972. Valley alluviation in southwestern Wisconsin. Annals of the Ass. American Geographers, vol. 62 (3): 401-410.

KNOX, J.C. 1975. Concept of the graded stream. In: Theories of landform development. Proc. 6th Ann. Geomorphology Symp., Binghamton, New York: 169-198.

KOLSTRUP, E. 1980. Climate and stratigraphy in northwestern Europe between 30'000 B.P. and 13'000 B.P., with special reference to The Netherlands. Med. Rijks Geologische Dienst, vol. 32-15: 181-253.

KOZARSKI, S. & K. ROTNICKI, 1977. Valley floors and changes of river channel patterns in the north Polish Plain during the Late Wurm and Holocene. Quat. Geogr. 4: 51-93.

KOZARSKI, S. 1983. River channel changes in the middle reach of the Warta valley, Great Poland Lowland. In: (S.Z. ROZYCKI, S. KOZARSKI, J.E. MOJSKI & L. STARKEL eds.) Quaternary Studies in Poland, 4, "Palaeohydrology of the Temperate zone": 159-169.

LEOPOLD, L.B. & M.G. WOLMAN, 1957. River channel patterns: braided meandering and straight. Prof. Pap. U.S. Geol. Survey, 282-B, 85 pp.

LEOPOLD, L.B., M.G. WOLMAN & J.P. MILLER, 1964. Fluvial processes in geomorphology. Freeman, San Francisco, 385 pp.

LEOPOLD, L.B. & M. MADDOCK, 1953. The hydraulic geometry of stream channels and some physiographic implications. U.S. Geol. Survey. Prof. Papers 252.

LOCKWOOD, J.G. 1983. Modelling climatic change. In: (K.J. GREGORY, ed.) Background to Palaeohydrology: A perspective. Wiley, Chichester, 35-50.

MOORE, P.D. 1983. Palynological evidence of human involvement in certain palaeological events. In: (S.Z. ROZYCKI, J.E. MOJSKI & L. STARKEL, eds.) Quaternary studies in Poland, 4, "Palaeohydrology of the Temperate zone": 97-105.

MULLENDERS, W. & F. GULLENTOPS, 1957. Palynologisch en geologisch onderzoek in de alluviale vlakte van de Dijle te Heverlee-Leuven. Agricultura 5: 57-64.

MULLENDERS, W., F. GULLENTOPS, J. LORENT, M. COREMANS & E. GILOT, 1956. Le remblaiement de la vallée de la Nethe. Acta Geographica Lovaniensia, 4: 169-182.

MUNAUT, A.V. & E. PAULISSEN, 1973. Evolution et paleo-ecologie de la vallée de la Petite Nethe au cours du post-Würm (Belgique). Annales Soc. Géol. Belgique, 96: 301-348.

ROSE, J., C. TURNER, G.R. COOPE & M.D. BRYAN, 1980. Channel changes in a lowland river catchment over the last 13'000 years. In: (R.A. CULLINGFORD, D.A. DAVIDSON & J. LENIN, eds.) Timescales in Geomorphology: 159-175.

SCHUMM, S.A. 1960. The shape of alluvial channels in relation to sediment type. U.S. Geol. Survey Prof. Paper 352-B.

SCHUMM, S.A. 1977. The fluvial system. Wiley, New York, 338 pp.

STARKEL, L. 1983. Progress of research in the IGCP-project No. 158, Subproject A. Fluvial environment. In: (S.Z. ROZYCKI et al., eds.) Quaternary studies in

Poland, 4, "Palaeohydrology of the Temperate zone": 9-18.

TAVERNIER, R. & G. DE MOOR, 1974. L'évolution du bassin de l'Escaut. In: (P. MACAR, ed.) L'évolution quaternaire des bassins fluviaux de la Mer du Nord meridionale, Liège: 159-233.

VANDENBERGHE, J. 1977. Geomorfologie van de Zuiderkempen. Verh. Kon. Acad. v. Wetenschappen, Letteren en Schone Kunsten van Belgie, Klasse der Wetenschappen, XXXVI, 3, 77 pp.

VANDENBERGHE, J., P. PARIS, C. KASSE, M. GOUMAN & L. BEYENS, 1984. Palaeomorphological and botanical evolution of small lowland valleys. A case study of the Mark valley in northern Belgium. Catena, vol. 11: 229-238.

VERBRUGGEN, C. 1971. Postglaciale landschapsgeschiedenis van zandig Vlaanderen. Botanische, ecologische en morfologische aspekten op basis van palynologisch onderzoek. Thesis. Gent, 440 pp.

VERBRUGGEN, C. 1979. Vegetational and palaeoecological history of the Late Glacial in Sandy Flanders. Acta Univ. Oul., A 82 3: 133-142.

Lake, Mire and River Environments, Lang & Schlüchter (eds)
© 1988 Balkema, Rotterdam. ISBN 90 6191 849 9

Evolution of the Morava river in Late Pleistocene and Holocene time

P.Havlicek
Geological Survey, Praha, Czechoslovakia

ABSTRACT: The stratigraphy and development of Late Pleistocene and Holocene deposits in the drainage area of Morava River was studied in detail during a geological and paleo-botanical investigation carried out in the Carpathian Foredeep and at the south-eastern border of the Bohemian Massif.

Detailed stratigraphical and geological studies of late Pleistocene and Holocene deposits in the Morava River drainage basin were part of the geological and paleobotanical investigation carried out in the Carpathian Foredeep and at the south-eastern margin of the Bohemian Massif. The Morava River flows throughout Moravia north-southward over a distance of about 300 km and begins its course on the Czechoslovak/Polish border at the foot of Mount Kralicky Sneznik and joins the Danube at Bratislava on the Czechoslovak/ Austrian border (at ca. 139 m a s l). Alluvial plains of the Morava, Hana, Becva, Dyje, Jihlava and Svratka Rivers were studied, respectively, by HAVLICEK (1977, 1980 a,b, 1983 a,b), OPRAVIL (1983), by ZEMAN (1968, 1973), RUZICKA (1968) and HAVLICEK & SVOBODOVA (1984).

The fluvial sandy gravel began to be laid down or redeposited in the Morava River alluvial plain at its confluence with the Dyje River during the Last Glaciation advance (as is indicated by radiocarbon data of 16'170 $\pm$ 480 y BP = Hv-9728 and 22'400 $\pm$ 3'650 y BP = Hv-7150) near Lanzhot village. Charcoal was found at the base of a Quaternary sequence about 40 m thick in the Straznice-Privoz locality; its age is 46'750 $\pm$ 3'940/2'630 y BP = Hv-9732.

Up to now there is no evidence, that the deposition and redeposition of the fluvial sandy gravel were accompanied by any major chronological break and it took place throughout Würm into middle, and locally, late Holocene time. During this period the

rivers often shifted their beds and erosion of various intensity alternated with depo-sition. Evidence used in support of this assumption is provided by discoveries made in fluvial sandy gravel of the Dyje and Morava alluvial plains. Based on radio-carbon data ranging from 7'990$\pm$ 75 y BP = Hv-9729 to 1'970 $\pm$ 80 y BP = Hv-9725, Neo-lithic and Lusatian pottery fragments and a medieval horseshoe were found, in addition to a blackish oak door frame at a depth of 6 to 8 m. Brick and pottery fragments dating from the 17th and 18th centuries were obtained from fluvial sandy gravel at two lovermost morphological levels of the Becva alluvial plain at Prerov, in addition to a small coin por-traying Francis the First in 1800. This gravel had a sparse vegetation cover and swampy assemblages were growing in erosion rills and depressions.

Blown sand dunes lay on the surface of the fluvial sandy gravel, especially in the confluence area of the Dyje and Morava. Wind-carved pebbles were found at the boundary between this gravel and aeolian sand, as in Mikulcice, indicating an intense aeolian activity in exposed alluvial plains mostly filled with river-deposited sandy gravel. On some dunes important centres of the Great Moravian Empire of the 6th to 10th century A.D. such as Mikulcice and Pohansko, including the Slavonic cementerey "Na Piskach" (9th century A.D.) at Dolni Vestonice in the Dyje River alluvial plain are recorded. Evidence available on the settlement in such alluvial plains comes especially from the middle

Burgwall period from about 1'000 to 1'150
y BP, as well as from the period characterized by a maximum settlement density in
the Dyje and Morava alluvial plains and
by the flourish and fall of the Great
Moravian Empire (Pohansko, Mikulcice,
Staré Mesto u Uherského Hradiste). These
important communities indicate, that at
that time river water levels were fairly
steady and alluvial plains were neither
heavily flooded nor attacked by groundwater for a long period of time. A
characteristic feature of these plains was
sparse vegetation consisting of oak, elm
and ash-tree, as can be seen at Straznice-
Privoz where 55 to 69 pieces of tree 50
to 500 years old grew per 1 hectare
(PRUDIC 1978). A subsequent increase in
flood activity and hence in groundwater
level caused the tree roots to decay and
whole trunks to become uprooted. The same
may probably be said of the extinction of
hornbeam in the middle basin of the Morava
at Staré Mesto u Uherského Hradiste prior
to the deposition of flood loam, i.e.
before the Great Moravian Empire came into
existence, some 1'150 to 1'050 y BP.

In the middle and especially late Holocene river beds began to change their
course and deposited flood loam beds at
various places and times. Thus, for
example, the flood loam at Veseli nad
Moravou was first laid down in a period
some 3'560 $\pm$ 130 y BP = Hv-7153 and that
known from Breclav in the Dyje alluvial
plain was laid down 3'720 $\pm$ 60 y BP =
Hv-9727, whereas the flood loam reported
from Straznice-Privoz was deposited in the
lower basin of the Morava much later:
850 $\pm$ 65 y BP = Hv-12'673.

Floods resulting in the deposition of
loam attained their peak toward the end
of the middle Burgwall, i.e. in a period
from about 750 to 1'000 y BP, and also in
the Middle Ages. Periods of flooding
alternated with those of sedimentation
tranquility to produce various soils on
the surface. Two to three subfossil soils
characteristically occurring in all the
alluvial plains of the Morava River basin
are known from a preserved sequence of
flood loam attaining an average thickness
of 3 to 4 m. Three subfossil soils can be
differentiated at Tovacov and Tlumacov in
the middle basin of the Morava River, of
which the middle soil has been dated
archeologically as coming from the 4th to
10th century of this era, i.e. from the
period of the migration of nations to the
fall of the Great Moravian Empire. There

was a decrase in the rate of erosion and
deposition during the Thirty Years War
associated with the formation of a young
subfossil soil. At Staré Mesto u Uherského
Hradiste, one of the most significant communities of the Great Moravian Empire,
flood plain sediments dating from the
8th century A.D. were found by archeo-
logists to lie at a depth of 2 to 4 m. A
cultural layer of the Empire overlying a
conspicuous subfossil soil suggests that
the soil ceased to form at the turn of the
9th and 10th century A.D. Subsequent to it
is a gap in sedimentation traceable till
1257 A.D., the year in which the community's
foundations were laid down. From the
mid-16th century onwards it is possible to
observe a sequence of cultural layers
based on archeological evidence and
indicating climatic changes in humidity of
various duration. A complex of flood plain
sediments containing at least two subfossil
soils was deposited on sandy gravel at
Straznice-Privoz in the Morava River
alluvial plain. A large number of coalified
tree trunks (ash-tree, elm, oak, etc.) were
discovered at the boundary between a lower
gley subfossil soil and still lower lying
fluvial sand; their age is estimated at
850 $\pm$ 65 y BP = Hv-12'673. At Pohansko,
specifically at the site called "Eastern
Gate", the Dyje River bed is abandoned
and filled with flood loam 1.70 m thick
together with sandy clay sampled for palynological studies by H. SVOBODOVA. According
to their results the sedimentation of the
flood loam began in the Lower Sub-Atlantic
period; the highest is the content of
Abies, synantropic species and of cultural
plants.

The section observed in the Jihlava
River bluff at the village of Ivan can be
described as follows: its first layer consists of sandy gravel lying in the Jihlava
alluvial plain; the second layer is black
subfossil soil containing bones of a
salmon associated with pottery coming from
the 9th to 10th century A.D. in its middle
part and from the 14th century A.D. at the
base of the next, third layer which
represents a complex of subfossil soils
and flood loam yielding Slavonic pottery.
The fourth layer consists of the youngest,
pale grey flood loam with recent alluvial
soil on its top. A similar section can
also be seen in the Svratka River meander
at Velké Nemcice village to contain flood
loam with 3 to 4 subfossil soils. Their
Upper Sub-Atlantic age is based on paleobotanical analyses by H. SVOBODOVA. The

Figure 1. Ivan-meander of Jihlava River: flood loam and subfossil soils, (Photo P. Havlicek).

Figure 2. Uherské Hradiste - Morava River: two subfossil soils from 8th (1) and 9th (2) century with intercalated flood loam; in the upper part "cultural sedimentation" after a hiatus until 1257 A.D. (3), (Photo P. Havlicek).

pollen diagram shows that the deposition of the complex of flood loams and sub-fossil soils began in the Upper Sub-Atlantic period that is characterized in this area by a deforestation of the landscape. Dominant synantropic and rural species Anthoceros punctatus and A.laevis and accessoric cultural plants (Fagopyrum) are present of the herbs. In the final phase of the deposition swamp and marsh species prevail.

Studies of the alluvial plains in the Morava River basin have shown that the youngest flood loam (High Middle Ages to Recent) perfectly smoothened their originally highly rugged surface and hence concealed most of their communities and cementeries dating from the Great Moravian Empire or still earlier periods. Areas unaffected by flood loam deposition are only covered with blown sand dunes.

REFERENCES

HAVLICEK, P., 1977: Radiokarbondatierung der Flussablagerungen in der Talaue des Flusses Morava (March). - Vest. Ustr. ust. geol., 52, 275-283, Praha.

HAVLICEK, P., 1980 a: Vyvoj terasového systému reky Moravy v hradistském prikopu. Anthropozoikum 13, 93-125, Praha.

HAVLICEK, P., 1980 b: Late Pleistocene and Holocene fluvial deposits in central and southern Moravia. Abstract of lectures of investigations outside the Main-Regnitz area, Universität Düsseldorf.

HAVLICEK, P., 1983 a: Late Pleistocene and Holocene Fluvial deposits of the Morava River (Czechoslovakia). - Geol. Jb., A 71, 209-217, Hannover.

HAVLICEK, P., KOCI, A., JANOSTIK, M. & A. SUTOR, 1983 b: Late Pleistocene and Holocene stream sediments in Moravia: New methods and results. - Quaternary studies in Poland, 4, 125-133, Warszawa-Poznan.

HAVLICEK, P. & H. SVOBODOVA, 1984: Excursion guide. Mikulcice Czechoslovakia 24-28 September 1984. IGCP Project 158 (subproject A)-Mikulcice.

OPRAVIL, E., 1983: Udolni niva v dobe hradistni. - Studie Archeol. ust, CSAV v Brne, Academia Praha.

PRUDIC, Z., 1978: Straznicky luh ve druhé polovine 1. tisicileti n. 1. - Lesnictvi, R. 24 (LI), Praha.

RUZICKA, M., 1968: Subrecentni sterkopiskova terasa na dolnim toku Becvy. - Vest. Ustr. ust. geol., R. XLIII, 5, Praha.

ZEMAN, A., 1968: Stratigrafie holocénu ve vychodni casti vyskovského uvalu. - Vest. Ustr. ust. geol., 2, R. XLIII, Praha.

ZEMAN, A., 1973: Pleistocenni fluvialni sedimenty Vyskovské brany. - Sbor. geol. Ved, R. A, 9, 45-76, Praha.

Lake, Mire and River Environments, Lang & Schlüchter (eds)
© *1988 Balkema, Rotterdam. ISBN 90 6191 849 9*

New methods of research on the Holocene palaeohydrography of Hungary

K.Mike
Research Centre for Water Resources Development, VITUKI, Budapest, Hungary

ABSTRACT: On the base of our investigations it might be said that in Hungary the genesis of lakes is closely related to the history of palaeo-rivers. During the Holocene, nevertheless, almost all of our large lakes became independent from the rivers which performed their own genesis, since the basins of lakes were formed already in the Pleistocene.

INTRODUCTION

About 2/3 of the area of recent Hungary is a filled-up flat-land, which means a subsiding terrain in relative terms. While in areas of uplift denudation is the main surface-forming factor, in the same time in subsiding areas accumulation will play the main role.

Our representative area is in the surroundings of the Tisza River (Fig. 1).

Holocene palaeohydrographical investigations (investigations of terraces, mineral- and grain-size composition, wearedness, roundness, radiocarbon, dendrochronological, palaeontological, etc. research and the informations given by drillings on certain points) did not yet give reliable answers in regards of lowland palaeohydrography, the events occurring in space and time and interrelations. By the use of the new method, developed by the aid of the Archeological Institute of the Hungarian Academy of Sciences as a mean by which the classical methods might become more complete, a more accurate approach became possible. In the followings these auxiliary methods will be explained.

METHODOLOGY

The level-correction relief-reconstruction method is able to reconstruct the former relief on the base of recent landsurface height data and the height-differences which occurred during the last 100 years (Fig. 2), using a correction (Fig. 3), according to certain dates (millenia). The hydrographic units have had to be situated along the deepest parts of the relief. Thus, the ancient beds of water-courses were traced.

The reconstruction of flood-plains might be first of all for archeological research. It is well-known that the level of floods is not a horizontal plane (like the surface of a lake) but has a slope. The levels of the preriver-draining floods were mapped, thus, the flood-levels might be trans-designed onto the reconstruction-maps too. By the aid of the reconstruction of flood-plains the areas covered perennially can be stated approximately, and the areas touched by extraordinary floods as well as the areas being always free of inundations (Fig. 4, 5).

By the aid of the geological slice-mapping: the geological formations of specified height above sea level are constructed (Fig. 6, 7). As a base the fact is used, that the plains of Hungary are explored according to a net of grid of drilled wells in every ca. 200 metres. (The drillings were carried-out for different reasons, but mainly for mapping purposes.) During the intense shifting of river-beds (ca. 1 km/100 years) always coarser sediments were deposited than in the surroundings, and according to a predetermined slope (evenly).

The maps of slicing method prepared in

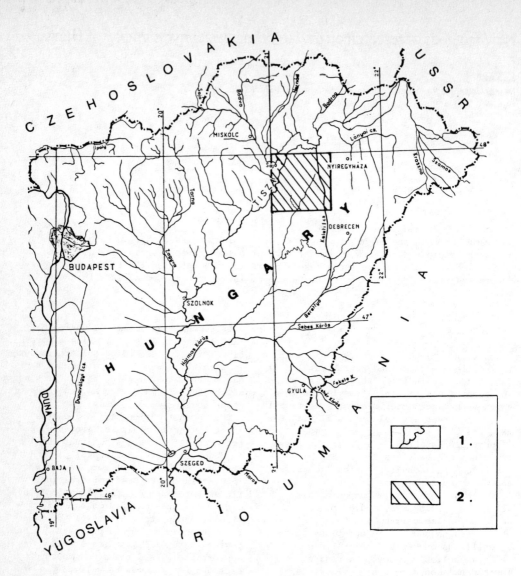

Figure 1. Index map for our study. 1 – recent hydrography, 2 – representative area.

every 2.5 metres do show the tracks of
water-courses in specified heights above
sea level, thus, the recent positions of
coarse sediments is given. But the question is: which sediment-layer belongs to
which, and from which time-span ?

On the base of airphoto interpretation
it was possible to identify and add more
accuracy to the results obtained by the
previous methods. Taking into account the
large-scale river-bed shiftings: along
the same strip probably rivers of differ-

ent discharges could have moved. The abandoned river-beds were different too. The
same river-course might have been in
different periods, or even in the same
time branches of different discharge.
Their identification can be done by airphotos. Sometimes the direction of flow
of a former river might become problematic
along a certain part of the river-bed. On
the base of the position of the natural
levees an answer might be got by the
traces of river-bed bends on airphotos.

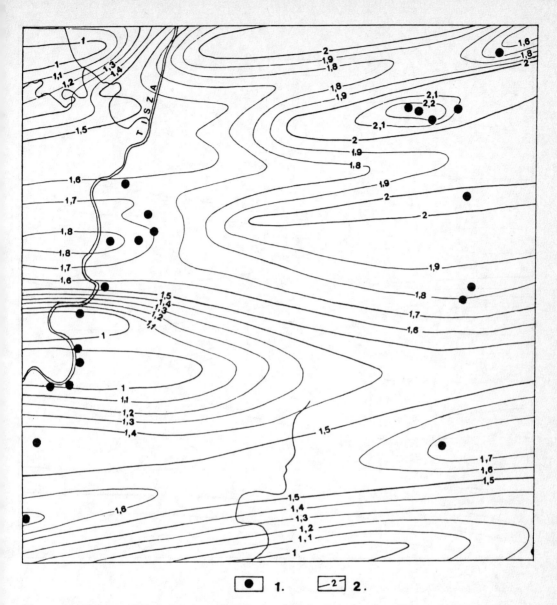

Figure 2. Changes of level (elevations). 1 - bench-mark for levelling, 2 - Isokinetic lines (mm/year).

SUMMARY

   By the application of these multi-method approaches the hydrographical picture within certain centuries can be reconstructed - and it can be further checked and made more accurate by conventional methods.

   In areas having different geological history: the reconstruction of palaeo-

hydrography might be done by using other methods. The palaeohydrographical changes explored by different means have to be put together like the particles of a mosaic - into a single sequence of events (either in terms of a country or a continent). On the base of our investigations it might be said that in Hungary the genesis of lakes is closely related to the history of palaeo-rivers. During the Holocene,

Figure 3. Isocorrection map for the last 4'900 years in m.

nevertheless, almost all of our large
lakes became independent from the rivers
which performed their own genesis. Since
the basins of lakes were formed already
in the Pleistocene.

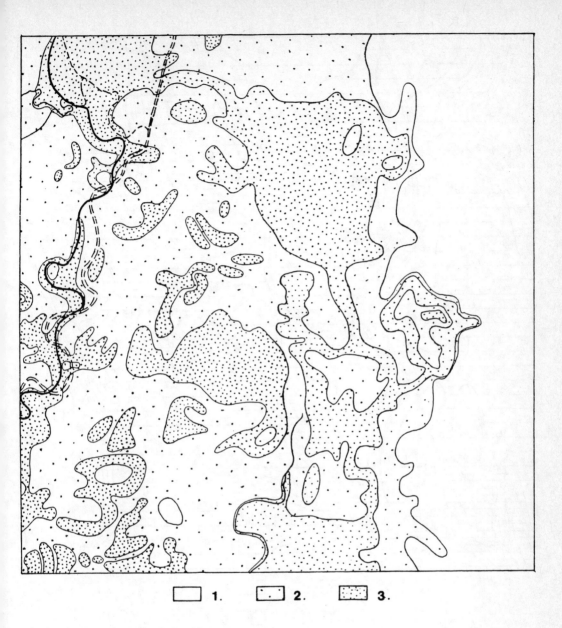

Figure 4. Pre-river-draining status. 1 - free of inundation, 2 - temporarily inundated area, 3 - perennially inundated area.

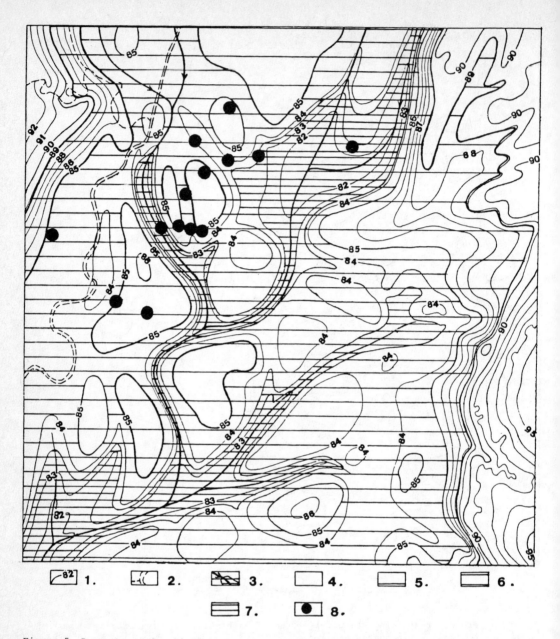

Figure 5. Reconstructed relief and hydrography status 4'900 years ago. 1 - reconstructed contour-line, 2 - recent hydrography, 3 - reconstructed hydrography, 4 - area free of inundation, 5 - area inundated only during flood-time, 6 - area inundated regularly in every year, 7 - perennially inundated area, 8 - archeological site.

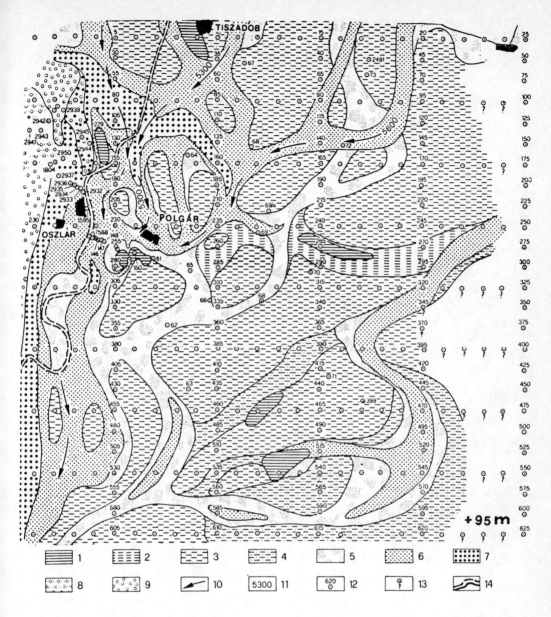

Figure 6. The geological formations of the flood-plain of the Tisza River, in an elevation of +90 metres above sea level (Baltic sea). 1 - clay, 2 - silt, 3 - aleurite, 4 - sandy aleurite, 5 - fine sand, 6 - sand, 7 - coarse sand, 8 - gravelly sand, 9 - gravel, 10 - flow-direction of ancient rivers, 11 - some flow directions of the base of the reconstruction of the relief, 12 - site and reference number of exploratory drillings, 13 - bore-hole which did not reach the desired level, 14 - the recent bed of the Tisza River.

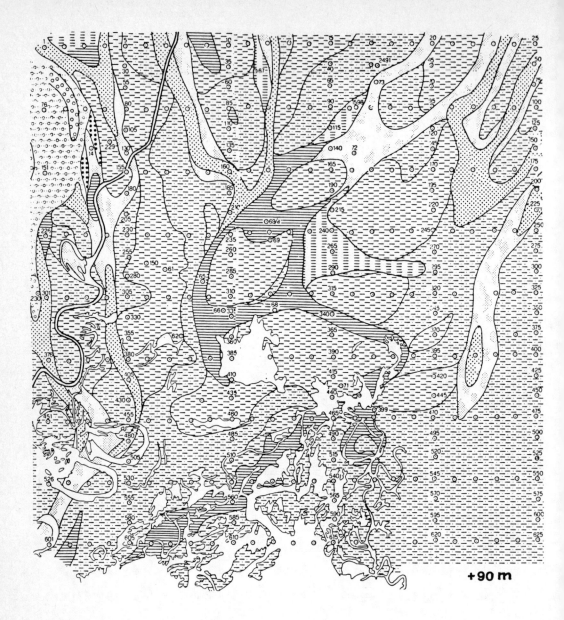

+90 m

Figure 7. The geological formations of the flood-plain of the Tisza River, in an elevation of + 85 metres above sea level.

Lake, Mire and River Environments, Lang & Schlüchter (eds)
© 1988 Balkema, Rotterdam. ISBN 90 6191 849 9

The history of the hydrography of Hungary and the Carpathian Basin determined by large-scale tectonics

Ö.Rádai
Research Centre for Water Resources Development, VITUKI, Budapest, Hungary

ABSTRACT: The hydrography of the Carpathian Basin and Hungary is a fascinating example for the importance of large-scale tectonic events on the physiographic evolution of an area. The earth's crust in Hungary (as it is well known from geophysical investigations) is exceptionally thin as compared to the average in other parts of the continents. The explanation might be given by the process of "subcrustal erosion", related to a so-called "geo-tumor", or more properly by the dynamics of several convection-cells. - The original, natural hydrography of the area was controlled by a system of straight-, curved-, and circular-elliptical-, concentric-annular lineaments displaying a very consistent pattern to which the recent hydrography is still strongly related to. Beyond the action of convection-cells several other explanations of geological events have to be taken into account as factors creating or "ornamenting" the hydrographical features.

INTRODUCTION

The special situation of the Carpathian Basin, in terms of the original hydrography and thermal-water occurrences, is determined by tectonism, basically by plate-tectonics.

The original hydrography, both surface- and subsurface drainage systems, existed until the beginning of the river-draining and swamp-draining activities in the second half of the last century. The aim of such developments was to eliminate the almost yearly inundation of vast, low-lying parts of the country and to gain land for agricultural purposes at the sites perennially and temporarily water-covered (like oxbow-lakes, abandoned river-bed sections, marshes, bogs and mires). Before these human activities, almost 2.3 million hectars were only used temporarily, because of periodical inundation. All that comes from the special situation of Hungary and the area surrounded by the Carpathian Mountain Ranges since Hungary is the recipient of waters draining the hills and mountains. The basin itself is drained by three major rivers: the Danube, the Tisza and the Dráva. The two latter discharge into the Danube too, but Beyond the borders of Hungary. - Though the major streams originate in the neighbouring countries, they enter an environment being totally different from the area which they come from. The Danube plus her right-hand tributaries originate in the German Federal Republic, Austria and Yugoslavia (a very small part of the water of the Danube comes from Switzerland) and the left-hand tributaries, the Tisza River and her tributaries enter the country from Czechoslovakia, the Soviet Union and Rumania (from parts lying within the Carpathian Basin).

OBSERVATIONS

The pattern of the surface water drainage system within the country is set by the streams entering from abroad - receiving the small water courses which originate in Transdanubia and North-Hungary and also the water transported by the ramified, today mainly artificial drainage systems and canals serving also other purposes in the plains.

From all the above it is quite evident that the drainage conditions of the country are (or at least were) poorly developed. The river-pattern in the plains is rather sparse and the slope of the water courses is extremely mild because of the clima-

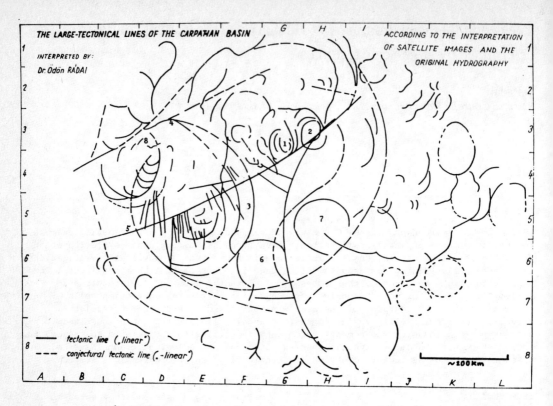

THE LARGE-TECTONICAL LINES OF THE CARPATHIAN BASIN

INTERPRETED BY:
Dr. Ödön RADAI

ACCORDING TO THE INTERPRETATION
OF SATELLITE IMAGES AND THE
ORIGINAL HYDROGRAPHY

—— tectonic line („linear")
--- conjectural tectonic line (.-linear")

~100 Km

Figure 1. 1 – Mátra Mountains, 2 – Bükk Mountains, 3 – Solt Plain, the eastern boundary of the formerly often flooded area along the Danube, 4 – The "focus" of the straight linears, 5 – Long, slightly curved linear feature preforming the southern side of the Lake Balaton and further of the Mátra- and Bükk Mountains, 6, 7 – Relatively small circular features in contact with large ones.

tological conditions and the very thick sediment layers covering huge parts of the country. Only a fragment of the total area might be considered as outcrops of solid rocks, obviously, this is the case in the hilly and mountainous areas.

The thickness of the sediment in the flat-land area is in several parts in the order of magnitude of 1000 meters. The fact that all such sediment traps contain deposits younger than Eocene indicates how fast the subsidence of the substratum took place. And thus we come to the special situation of the Carpathian Basin and to the question: how to explain the extra-ordinary features related to it?

Two main characteristics of the area are known already since the last century. One is the fact that this area is free, or al-most free of devastating earthquakes (in spite of the relatively frequent occurrence and severeness of such events in the sur-

roundings, e.g. in Yugoslavia and Rumania). The other factor is given by the special hydrographic situation with an abundance of thermalwater under the flat areas and young sediments being explored and ex-ploited by hundreds of wells.

Soon, by the use of up-to-date geo-physical methods, it became clear that the earth's crust is exceptionally thin within the arch of the Carpathian Ranges; being only about 18 kilometers thick to the con-trary of the about 30 kilometers of the crust of the continents (STEGENA-HORVATH 1978). The phenomenon must have been in close connection with the concept of plate tectonics (STEGENA-GECZY 1975). Also, evidences for geomechanical forces acting in a compressive mode from SE and pushing against an "obstacle" in the opposite direction (NW) surprisingly not along one single line but shifted apart are postu-lated (RADAI 1968). These formed very

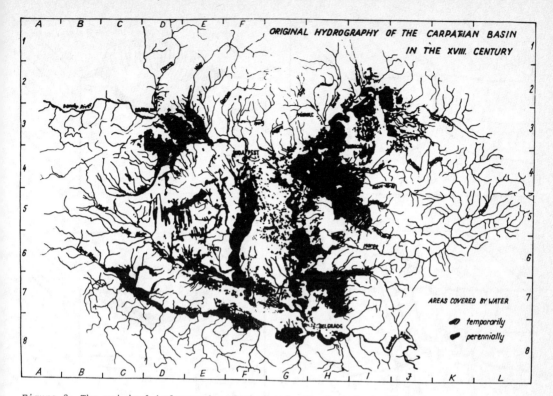

Figure 2. The original hydrography of the basin displays huge flooded areas.

peculiar features, e.g. so-called parquette-like joint-fault system (RADAI 1962).

After the hydrogeological interpretation of thousands of airphotos (covering the mountainous part of the country) it was recognised that bent-arcuate lines and concentric patterns of lineaments are to be found frequently in these regions. All these forms were of "meso" dimensions – being only parts of mountains (e.g. within the karstified rock-formations of the Bükk Mountains in NE Hungary, RADAI 1978b). But some remarkable traces of much more extensive lineaments were recognised as well: curved tectonic lines transsecting the whole mountain. On the base of airphoto interpretation these results seemed to be preliminary, because of the scale (about 1:25'000) of the available photos and because of the large number of airphotos which were necessary for the coverage of only parts of the mountain ranges. Nevertheless, the existence of curved- and circular (elliptical) lines became apparent, though only fragments could be seen.

A new era of study was initiated in Hungary: the intense application of air-photos and of remote-sensing methods in general (RADAI 1969). The methods were mainly used for deciding on suitable sites for karstwater-exploration and exploitation wells. But soon the results became an integral part of the construction of maps for other hydrogeological purposes too – like the geological basement maps, or for the delination of protective areas of thermal springs and the construction of karstwater-table maps in areas strongly influenced by the lowering of the karstwater-level in connection with mining activities, etc. And then the epoch of space imagery brought the long-desired opportunity: to see huge parts of the country in one or a few pictures (LANDSAT). Immediately it became obvious: the entire actual hydrography of the Carpathian Basin is almost totally controlled by a large-tectonic system. The lines being traceable on space images showed two main groups: straight ones and curved, circular-elliptical, concentric and annular line-systems (RADAI 1978a).

The strong dependency of the observed lineaments and the hydrographic system is striking in the first period of the study

147

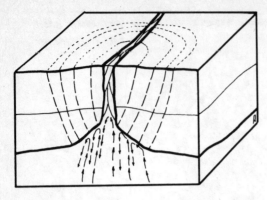

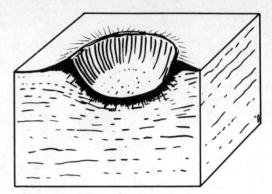

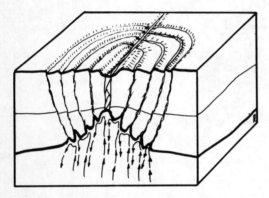

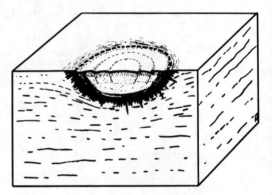

Figure 3 (above) - Figure 4 (below).
"Candelabrum" pattern: two stages of its
development, created by a convection-cell
in the mantle, eroding along a straight
and deep-reaching tectonic line (often a
horizontal fault).

Figure 6 (above) - Figure 7 (below).
Two stages in the formation of a circular
feature: impact of a meteorite or asteroid
and the erosion plus filling-up and subsidence of the infill.

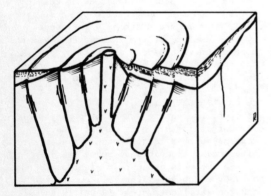

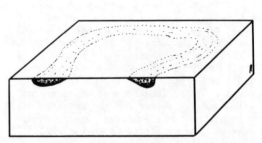

Figure 5. Volcanic form: caldera produced
by the subsidence of formerly melted
material.

Figure 8. Ring-like feature, formed by a
filled-up river bed.

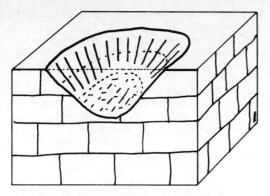

Figure 9. Doline (funnel-shaped feature in solid, soluble rock, e.g. in limestone) might be also filled-up, but the dimensions are relatively small - in the order of magnitude of tens, or hundreds of metres.

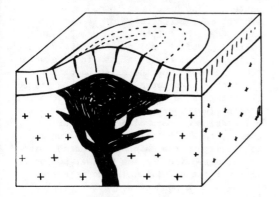

Figure 10. Intrusion of melted rock forms a "dome" and an annular pattern of joints.

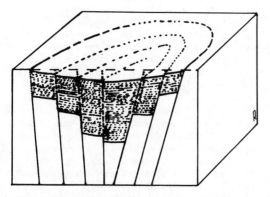

Figure 11. Typical depression, filled-up and subsiding because of compaction of the infill.

some of the experts actually denied the existence of the linear features. But other evidence could also be collected, supporting the really astonishing system (RADAI 1978b). Among these evidences were: water-chemistry data with differences in composition in neighbouring wells of the same depth and penetrating into the same geological formations, but one being in contact with a deep-reaching fault and the other in solid rock (the fault giving the possibility for groundwater to rise from formations which were separated from the rock-masses into which both wells were drilled). Age-determinations on subsurface water brought similar results than the above mentioned chemical data. Water-level oscillations, caused by earthquakes far away, to be observed in one well (in touch with a lineament) and not in another close to it, are another prove for the existence of the linear feature to be traceable only in satellite imagery. The most obvious evidence is still simply the existence of mountains and the hard-rock base of filled-up basins, being built of the same rock-type and with vertical displacements of several hundreds of meters while being very close horizontally.

Soon, the hope to have additional, exact evidence will be met too: a system of geodaetic fix-points was established. Their position was determined already and will be redone with great accuracy by most up-to-date surveying methods, in terms of distances, elevation and horizontal angles. The fix-points are attached to rock-surfaces of two neighbouring rock-blocks being separated by a tectonic line which dissects a part of the Transdanubian Mountain Ranges (RADAI 1976a). The site was chosen on the base of careful satellite-image interpretation, airphoto-interpretation and field-work (including information on geology, hydrogeology, geophysics and hydrography).

As it is to be seen in Fig. 1 the hydrography of the Carpathian Basin is closely related to the large-tectonic system of the area. But it has to be mentioned that a very special "tool" was used for proving that connection: the map showing the natural hydrography of the area. That is the XVIIIth century status, without the man-made changes of today. It shows the already mentioned vast areas of perennially- or at least temporarily inundated land, and since, this situation is almost totally transformed by river-draining and land-reclamation plus canalization works.

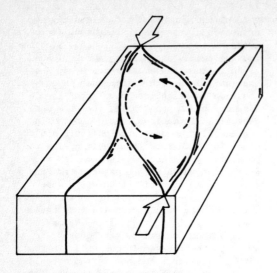

Figure 12. Two forces acting against each-other, but not along the same line; the result is an almost almond-shaped feature with a center part rotating and a very sophisticated pattern surrounding it.

INTERPRETATION

Already in the earliest stages of these investigations attempts were made for explaining the observed tectonic phenomena. First of all the "geotumor", or "subcrustal erosion" concept (STEGENA-HORVATH 1978) was considered and the very early finding of MOLDVAY (1969).

According to the geotumor concept the exceptionally thin crust was produced by subcrustal erosion caused by a large convection-cell in the earth's mantle. This situation is "responsible" also for the positive geo-thermal anomaly making Hungary so rich in hot water springs and drilled wells.

Beyond the geotumor theory the observations had to be evaluated in the context of rather peculiar features "ornamenting" the large-tectonic image of the Carpathian-Basin: features like rigid-straight linears "radiating" from the point where the Danube enters Hungary (border of Austria and Czechoslovakia). But, elliptic forms (e.g. those marking the Mátra Mountains) needed apparently different explanations. Very strange were some water-courses where tributaries joined in the form of multi-armed candlesticks (candelabra, Figs. 3, 4).

Not uplifted round blocks (called "diapiroides") were discernible only, but also

subsiding ones, which are referred to as "depressions" (Fig. 2).

The straight-rigid lines are the results of subcontinent-size plate-tectonic movements, with either compressive, or dilatating character and acting parallel but shifted apart along the two characteristic directions, NW-SE. - The Hungarian Mountain Ranges were formed by compressive forces and lifted up. The dilatating forces are more likely to be responsible for the tectonic lines transsecting the ranges in more-or-less perpendicular directions.

The elliptical - annular form of the Mátra Mountains are clearly the result of a collapsing caldera. The area is of volcanic origin and the existence of a caldera was proposed a long time ago, its form and dimensions were detectable for the first time on the LANDSAT images (Fig. 5).

The multi-armed "candelabrum" drainage pattern might have been formed by subcrustal convection-cells of lesser dimensions than that of the "geotumor". These cells ("geoblisters") are relatively mobile, at least along the straight lines to which they are bound. By the shifting of the blisters a candelabra-series can be formed. The stem of the "candlestick" itself being a main river-course and the "arms" are the tributaries on both sides. Interesting enough, the tributaries might be running parallelly to the main water-course, either in the direction of flow or just against it - until they will join the main river-course with a bend of a quarter circle (Figs. 3, 4). - Of course, alternative explanations for the formation of these phenomena might be given: volcanic activity (Fig. 5) such as crater formation, the impact of asteroids (Figs. 6, 7), meteorites and intrusions (laccolithes, batholites) would be the most obvious ones (Fig. 10). - Dissolution, e.g. dissolving of limestone of gypsum, rock-salt, etc. by water and producing sink-holes, might also result in the formation of similar features (Fig. 9). - Circular forms are also common as atolls (coral rings built on top of slowly subsiding conical volcanic hills in the sea, according to the theory of CH. DARWIN). Curved lines are often the result of either erosion by water currents (Fig. 8) or of sedimentation, such as natural levees along river-banks and sand bars in the sea; another genetic possibility are dunes (see also Figs. 11 and 12, depressions and rotating rock masses).

SUMMARY

The actual and of course the original,
natural hydrography of the Carpathian
Basin and of Hungary is strongly influenced
or in most cases totally controlled by the
large scale tectonic activity of the area.
The tectonic pattern of this part of
Europe is mainly the result of a geotumor
(subcrustal erosion caused by the latent-
flowing material of the upper mantle of
the earth). Other factors, like relatively
small-scale convection-cells (geo-blisters)
of the mantle-material, impact of asteroids
and meteorites, dissolution of rocks,
deposition, action of organic bodies
(corals?) and last-but-not-least plate-
tectonics, might be responsible for the
"ornamenting" forms which give the characteristic pattern of the hydrography of the
area.

All these forms, being very consistent
and certainly "logical", were revealed by
the interpretation of thousands of airphotos and several satellite images; the
investigations have been supplemented by
the maps with reconstructions of the
original, natural hydrography of the area.

REFERENCES

MOLDVAY, L. 1969, 1970. Features of neo-
  tectonical surface-development in the
  Hungarian Mountain-Ranges (in Hungarian).
  I. MAFI EVI JELENTES 1969: 587-637.
  Budapest, II. MAFI EVI JELENTES 1970:
  155-179. Budapest.
STEGENA, L. & B. GECZI 1975. The Late-Caino-
  zoic development of the Pannonian Basin
  (in Hungarian).
  Földtani Közlöny: 105/(2): 101-123.
  Budapest.
STEGENA, L. & F. HORVATH, 1978. Critical
  tectonics in the Thetys and Pannonian
  epochs (in Hungarian).
  Földtani Közlöny 108: 149-157. Budapest.
RADAI, Ö. 1968. Use of photogrammetric
  interpretation for karstic water research
  in Hungary. Hungarian Reports 11th International Congress of the ISP, Lausanne,
  p. 71-76, Lausanne.
RADAI, Ö. 1969. Aerophotographic interpretation and karsthydrological mapping
  of karstic areas. VITUKI Papers and
  Research Results Vol. 28: 82 + 63 p.,
  100 Figs., 62 Photos, 2 Maps. Budapest.
RADAI, Ö. 1976a. The detection of recent
  crustal movements by aerophoto-inter-
  pretation. Proceedings of the XIIIth
Congress of ISP, 1976, 6 p., 5 Figs.,
  Helsinki.
RADAI, Ö. 1978a. Subsurface water environment and the reconnaissance of it by
  aerospace methods in Hungary.
  Proc. of the ISP - IUFRO Symp. Freiburg,
  13 p., 7 Figs.. Freiburg, GFR.
RADAI, Ö. 1978b. Environmental protection
  of karstwater by the interpretation of
  aerospace images.
  Proc. Int. Karsthydrological Symp.
  Budapest, 8 p., 5 Figs.. Budapest.

Lake, Mire and River Environments, Lang & Schlüchter (eds)
© 1988 Balkema, Rotterdam. ISBN 90 6191 849 9

Holocene valley development on the Upper Rhine and Main

W.Schirmer
Abteilung Geologie der Universität, Düsseldorf, FR Germany

ABSTRACT: The results on the Upper Rhine presented here, compared to those on other
Central European Rivers, give certainty that the reworking phases of the River Main
cannot be of local nature. There is a major control - the Holocene climate with its
fluctuations - effecting the rhythmics of reworking phases in the river valleys.

1 AIMS

On the Main River a detailed sequence of
Late Glacial and Holocene reworking phases
of the river has been worked out (Fig. 1,
and SCHIRMER 1983b). The question was,
whether this sequence of reworking phases
would be local - caused by climatic events
and human impact on the valley - or would
indicate an influence of widespread climatic control. The latter has been concluded from the following facts: (1) The
terraces of the Late Glacial Period have
undoubtedly been effected by distinct and
well-known climatic changes, as for
example the Younger Dryas Period. (2) Style
and structure of the Holocene terrace range
continue exactly those of the Late Glacial
Period. (3) Man's impact becomes evident
by quite other features, e.g. flattening
of the river channel bottom combined with
a tendency towards braiding - as it is the
case in the Unterbrunn and Staffelbach
Terrace along the River Main. Further
traces of man's influence are:
- augmentation of flood sediment
- input of preweathered soil material into
  the valley floor by soil erosion processes in the catchment area of the
  river
- increase of carbon content in the floodplain sediments
- change of the floodplain vegetation by
  clearance activities in and outside the
  valley floor. This effects a change of
  the wood spectra, preserved within the
  terrace gravel, as well as a change of
  the pollen spectra of the flood sedi-

ments
- increase of anthropogene finds in the
  channel sediment. (4) On the rivers in
Middle Europe a terrace sequence comparable with that of the River Main did
not exist up to now. But there exists a
lot of local studies (reviewed in
SCHIRMER 1973, 1974) and, moreover, own
local unpublished studies along the rivers
Saar, Mosel, Erft, Lahn, Sieg, Ruhr and
Lippe (all belonging to the Rhine catchment area), which all fit well into the
system of reworking phases of the River
Main.

Nevertheless further and more detailed
evidences are necessary to give the
proof that climate would have caused
different rivers to rework the valley
sediments simultaneously and rhythmically.
Thus, the following investigation of the
Upper Rhine area north of Strassburg
(Strasbourg) may give an example to serve
this aim.

2 HOLOCENE VALLEY DEVELOPMENT ON THE RIVER RHINE NORTH OF STRASBOURG (STRASSBURG)

2.1 Landscape zonation

In the Alsace 25 km north of Strasbourg,
the westward valley plain of the Rhine
River represents an inner part of the
Upper Rhine Graben. There, a cross section
of the valley surface shows a distinct
morphological zonation of the floodplain
landscape. This morphological zonation is

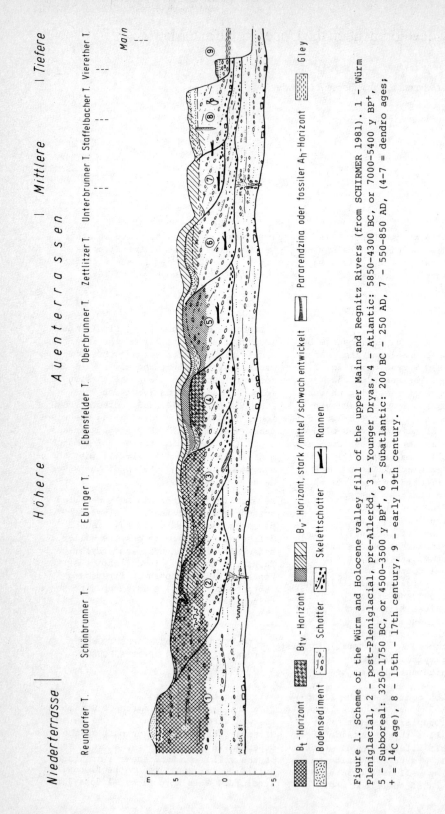

Figure 1. Scheme of the Würm and Holocene valley fill of the upper Main and Regnitz Rivers (from SCHIRMER 1981). 1 – Würm Pleniglacial, 2 – post-Pleniglacial, pre-Alleröd, 3 – Younger Dryas, 4 – Atlantic: 5850–4300 BC, or 7000–5400 y BP+, 5 – Subboreal: 3250–1750 BC, or 4500–3500 y BP+, 6 – Subatlantic: 200 BC – 250 AD, 7 – 550–850 AD, (4–7 = dendro ages; + = 14C age), 8 – 15th – 17th century, 9 – early 19th century.

accompanied by typical field names under-lining the character of the morphological zones as ecological zones (SCHIRMER 1985).

The Rhine is bordered immediately by the Auwald Zone (Riverine Forest Zone). Nowadays this zone is almost completely wooded and diked.

To the west follows the Wörth Zone ("Wörth" means an island within a river). This landscape is occupied by agricultural fields, and more rarely, by meadows. It was flooded by all major floods of the Rhine up to some decades ago. Well pre-served large meander bows of relatively young age mark this zone (e.g. in the area of Sandwörth, Altwörth and Neuwörth near Offendorf and in Mühlwörth near Gambsheim; see Fig. 2).

Towards the valley edge there follows the Felder Zone (Field Zone). It takes the highest position within this side of the Rhine valley floor (e.g. Bühlfeld, Stockfeld and Neufeld around Gambsheim and Offendorf). Old settlements of the valley floor followed this zone - which can be derived from the ending -heim of the place names - as did the lines of the old main roads and of the railway forming the Alsatian north-south connection. A well developed morphological subdivision of this zone does not, however, deny its homogeneous general character. To the west there follows the Matten Zone (Meadow Zone). This plain, smoothly de-scending towards the valley edge, was formerly a pure meadow landscape, as suggested by the field names "Hutmatt", "Riedmatten", "Katzenmättel" and "Boden-matt" situated between Gambsheim and Herrlisheim. But nowadays it is used as arable land. There also occur small areas of forests, and in wet depressions reed vegetation. This zone represents a morpho-logical transition from the Felder Zone to the Ried Zone. The outmost part of the valley floor of the River Rhine is formed by the Ried Zone (Reed Zone). This zone borders the Tertiary hill country which is higher in elevation.

2.2 Geological zonation

Geological and pedological investi-gations made by SCHIRMER and later on by SCHIRMER & STRIEDTER (1985) revealed these landscape zones to be a result of a fluctuating river history. Each zone represents a particular sedimentation period or several sedimentation periods; the boundaries of the landscape zones are also sedimentary unit boundaries.

Auwald Zone

The early topographical map of TULLA (1938) shows this zone lying eastward of Offendorf completely in the area of the branching network of the Rhine as present before river correction.

Thus, TULLA's map shows the Auwald Zone being an area of active fluvial working in the early 19th century. The gravel of this zone contains young cultural finds, e.g. near Fort Louis (GEISSERT et al. 1976), situated down the river in map Fig. 2. At locality 1 in Fig. 2 rannen (fossil tree trunks) occur with preserved cutting traces at the base of the trunk.

In the gravel pit Offendorf-"Fahrkopf" (locality 1) a 2 m section above the groundwater-table is exposed. Above the 19th century gravel there follows 60 cm of fine sandy flood sediment (Fig. 3). On top of it, a 10 cm thick calcareous pararendzina has been developed. The lime content of the C horizon is 19-23 %. In the A horizon it diminishes from the base to the top to 14 % - evidence of weak decalcification of the pararendzina.

Wörth Zone

The floodplain terrace of the Wörth Zone is 1-2 m higher in elevation than the Auwald level. Its surface has an undulating fresh morphology. A fair number of large meander bows has been well pre-served, and are now partly filled with groundwater. The gravel of the pit Gambsheim-"Kälbergrün" contained a mill-stone, and that of the gravel pit Offen-dorf-"Sandwörth" ceramics as well as rannen with preserved cutting traces.

The cultural finds date the gravel from the High Middle Ages on.

There is a distinct hiatus evident between the Wörth Zone and the Auwald-Zone, consequently the Wörth Zone gravel has to be assigned to the older part of the time span.

The gravel is covered by a 0.5-1 m thick fine sandy floodplain sediment. Its soil profile exhibits a 25 cm thick calcareous humus at its top, followed below by a B horizon of 20-25 cm thickness, weakly brown and slightly decalcified (B hor-izon 12 % $CaCO_3$; C horizon 18 % $CaCO_3$). The soil type is a braunified pararendzina.

A great number of older rannen lies in deeper gravel beds below that of the Wörth Terrace, as well as below gravels of the Auwald Terrace. They give evidence that both gravels, Wörth and Auwald, are super-posed over different gravels of older Holocene age.

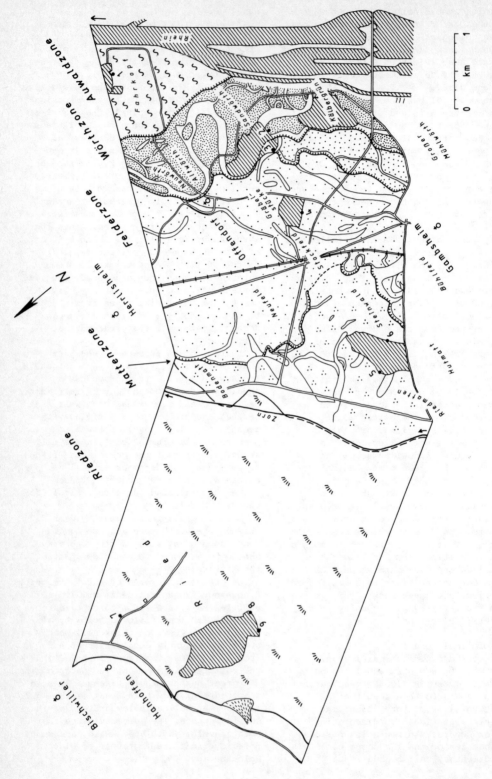

Figure 2. Terrace sequence of the Rhine plain 25 km north of Strasbourg (Strassburg), from SCHIRMER & STRIEDTER 1985.

156

Felder Zone

Fig. 2 shows this zone subdivided into two distinct terraces. Farer to the north K. STRIEDTER (in preparation) could meanwhile subdivide this zone into three terraces.

Lower Terrace of the Felder Zone
(Densly dotted small area in Fig. 2)

Their mean elevations lie about 0.5 m above the Wörth Zone surface. The gravel dates to the Early Middle Ages or older. The former date is indicated by the trunk of a ranne with the final tree ring year 589 AD, excavated in this terrace area in the gravel pit Offendorf-"Sandwörth". Rannen of this age occur frequently in this region (BECKER 1982:45).

At locality no. 3, in the upper terrace section the gravel is overlain by a fine sandy floodloam sediment up to 1.4 m thick. Below the humus horizon (30 cm thick) a distinct B_V horizon of 35 cm thickness has developed. The A horizon is rather decalcified (A horizon: 5 %, B: 14 %, C: 19 %). Soil type: pararendzina-brownearth.

Higher Terrace of the Felder Zone
(Widely dotted large area in Fig. 2)

This highest part of the floodplain landscape in the Rhine plain rises up to 1 m higher than the Lower Terrace of the Felder Zone. Under the groundwater-table of the gravel pit Gambsheim-"Gräbelstücke" there occur three rannen layers (R 1-3) dating to the middle and late Subboreal (^{14}C dating by M. GEYH, Hannover).

R1 8 m bel. surf. 3030+55 y BP (Hv-13 118)
R2 12 m bel. surf.
R3 17 m bel. surf. 3600+55 y BP (Hv-13 117)

Above the gravel of this terrace body the fine sandy floodloam of 60 cm thickness is completely decalcified and altered to a brownearth; the weathering also affects the upper parts of the underlying gravel (see Fig. 3).

Matten Zone

Towards the valley edge the surface of this zone gradually descends down to 1 m below the Felder Zone. Two ^{14}C dates of oak rannen date the gravel of the pit Gambsheim-"Steinwald" to the middle up to the late Atlantic period.

top of the gravel   4430+65 y BP (Hv-13 114)
deeper part of the
  gravel below the
  water level   5810+70 y BP (Hv-13 116)

The former soil on the fine sandy-silty flood sediment was a black pseudo-chernozem which, however, later on has been transformed to a parabrownearth. Both soils are now completely decalcified.

Ried Zone

Descending to the valley edge down to 1-3 m below the Matten Zone, the Ried Zone occupies the deepest position within the valley cross section. No wonder that the groundwater-table lies near the surface and the floodplain is overgrown with reeds. The upper parts of the gravel in the pit Hanhoffen are free of finds. This hints at a pre-Holocene age of the gravel. As for its deep position in the valley it should be of Late Würmian age. This age is confirmed by a pollen diagram from peaty parts of the Ried analysed by HATT (1937:70) the base of which indicates a Late Würm age.

On higher positions of that zone, on top of a 90 cm thick silty-fine sandy flood loam, a stony parabrownearth was developed. Later on, it was buried by a 70 cm thick fine sandy flood loam that includes ceramic finds. The surface of this loam is marked by a brownearth soil profile.

Deeper positions of that zone - the typical reed landscape - are completely gleyed. All such sections expose the surface soil as a wet-gley (Nassgley) or a gleyed brownearth, followed below by a second, buried, wet-gley. Pollen analyses (U. ERTL, Düsseldorf) date this buried soil containing Fagus, Secale and Plantago lanceolata to be subboreal or younger.

The geological map (scale 1:25'000, sheet Brumath-Drusenheim) of this region (MAIRE et al. 1972) shows the Holocene terraces and gravel bodies of the Matten Zone, Felder Zone and even part of the Wörth Zone as an area of Würmian gravel aggradation. The small seam along the Rhine River left in the map as Holocene sand and gravel alluvium actually corresponds to only the youngest Holocene, from the Late Middle Ages on.

While the Rhine plain has formerly been interpreted as a result mainly of Würmian sedimentation, the investigations of SCHIRMER & STRIEDTER (1985) give evidence that the plain is the result both of Late Würmian sedimentation and, to a large degree, periodically gravel reworking in the Holocene.

Meanwhile K. STRIEDTER is continuing this study upstream and downstream of our research area - an investigation for taking his doctor's degree. He added to

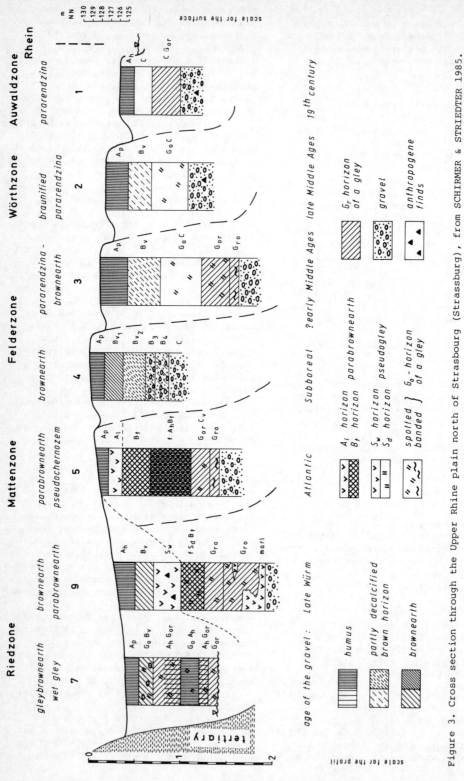

Figure 3. Cross section through the Upper Rhine plain north of Strasbourg (Strassburg), from SCHIRMER & STRIEDTER 1985.

158

the two terraces of the Felder Zone a third one, thus, subdividing the Felder Zone into a Higher (Subboreal), Middle (Roman Period) and a Lower one (Early Middle Ages) (K. STRIEDTER, in preparation).

3 MAIN AND UPPER RHINE

These preliminary results from the Alsatian Upper Rhine give evidence that the Holocene river history in the Upper Rhine Graben (with its tectonic subsidence) included also periodic events of sedimentary activity, as it is known from other regions such as the Alpine foreland, the Central uplands and the Northern German lowlands, and, in great detail, from the River Main (SCHIRMER 1983).

As on the River Main, the Holocene reworking phases of the Upper Rhine are represented as row terraces (SCHIRMER 1980:13, 1983a:201, 1983b:28). Likewise these row terraces can be well separated by morphological and pedological criteria (cf. SCHIRMER 1983a,b).

Up to now the dating of the individual terrace bodies on the Upper Rhine has not been accomplished in a similar level of detail as on the River Main. Therefore a stratigraphic correlation of both rivers can only be done roughly. However, all six Holocene terraces known from the Alsace so far have, according to their first datings, good equivalents in the Holocene terrace sequence of the River Main (compare Fig. 1, 3).

Moreover, the soil development also exhibits similarities. As the lime content of the unweathered substratum is much higher on the Upper Rhine (about 20 %) than on the River Main (about 0.5 %), the soil development is not identical. However, there is evidence of a predominance of black floodplain soils of the pseudo-chernozem type in the Late Würm and older Holocene - on the Upper Rhine up to the late Atlantic Period, on the River Main only to the Preboreal. It is followed or replaced by the development of a para-brownearth - in both valleys upon the flood sediment of the Atlantic gravel at the latest. The terraces from the Sub-boreal on exhibit soils of the developmental range pararendzina to brownearth, from the youngest to the older terrace, with a gradual intensification of development (see Fig. 3). This soil development, of course, proceeds much more quickly in the Main terrace sequence with its low lime content than in the Upper Rhine terrace sequence with its higher lime content. Thus, the Main sequence exhibits the brownearth stage in external parts of the Staffelbach Terrace (SCHIRMER 1983b:24), which correspond approximately to the Wörth Terrace of the Upper Rhine. In contrast, the Wörth Terrace only exhibits the incipient development of brownification.

Despite the influence of parent material composition, it is nevertheless clear that time is the dominant control on soil profile development from the Würmian Period on up to today. This enables to date floodplain terraces by their soil - provided that the soil catena has been adjusted to the age of the respective terraces at a proper place.

A further attempt to prove this stratigraphical system of reworking phases are the investigations on the Lower Isar River and adjacent Danube River made by G. SCHELLMANN (in preparation) on the base of former investigations by BRUNNACKER (1959) and HOFMANN (1973). There is a quite similar range of floodplain terraces, coinciding stratigraphically with that of the River Main. There is also a well-differentiated soil catena in which each soil type marks a distinct floodplain terrace. Due to a lime content of the floodplain sediments, much higher than that on the Upper Rhine, the Holocene soil catena moves within the range pararendzina to pseudo-chernozem.

The first results on the Upper Rhine presented here, together with those on other Central European rivers, give certainty that the reworking phases of the River Main cannot be of local nature. There is a major control - the Holocene climate with its fluctuations - effecting the rhythmics of reworking phases in the river valleys. Consequently, the reworking phases of the River Main cannot be restricted to Middle Europe. They should occur much more spread out, modified locally by climatic features of an individual region, and, in the younger Holocene Period, modified by man's activity of that region.

4 ACKNOWLEDGEMENT

Thanks to Dr. R. Brakenridge, Dartmouth College Hanover N.H., for reviewing parts of this text.

5 REFERENCES

BECKER, B., 1982: Dendrochronologie und Paläoökologie subfossiler Baumstämme aus Flussablagerungen. Ein Beitrag zur nacheiszeitlichen Auenentwicklung im südlichen Mitteleuropa. - Mitt. Komm. Quartärforsch. österr. Akad. Wiss., 5, 120 S., Wien.

BRUNNACKER, K., 1959: Zur Kenntnis des Spät- und Postglazials in Bayern. - Geologica Bavarica, 43: 74-150, München.

GEISSERT, F., MENNILET, F. & G. FARJANEL, 1976: Les alluvions rhénanes plio-quaternaires dans le département du Bas-Rhin. - Sciences géol., Bull., 29 (2): 121-170, Strasbourg.

HATT, J.P., 1937: Contribution à l'analyse pollinique des tourbières du Nord-Est de la France. - Bull.Serv.Carte géol. Alsace-Lorraine, 4: 1-76, Strasbourg.

HOFFMANN, B., 1973: mit Beiträgen von BADER, K., GANSS, O. & J.-P. WROBEL: Geologische Karte von Bayern 1:25'000, Erläuterungen zum Blatt Nr. 7439 Landshut Ost. - 113 S., 7 Taf., 2 Beil., München.

MAIRE, G., CLOOTS, A.R. & J.G. BLANALT, 1972: mit Beiträgen von SITTLER, C., GEISSERT, F., HUMMEL, P., THEVENIN, A., STIEBER, A. & P. SCHWOERER: Notice explicative de la Carte géol. de la France 1:50'000, no 234, Brumath-Drusenheim, Orléans.

SCHIRMER, W., 1973: State of research on the Quaternary of the Federal Republic of Germany. C 2. The Holocene of the former periglacial areas. - Eiszeitalter und Gegenwart, 23/24: 306-320, Oehringen/Württ.

SCHIRMER, W., 1974: Holozäne Ablagerungen in den Flusstälern. - In: WOLDSTEDT, P. & K. DUPHORN: Norddeutschland und angrenzende Gebiete im Eiszeitalter: 351-365, Stuttgart (Koehler).

SCHIRMER, W., 1980: mit Beiträgen von BECKER, B., ERTL, U., HABBE, K.A., HAUSER, G., KAMPMANN, T. & J. SCHNITZLER: Exkursionsführer zum Symposium Franken: Holozäne Talentwicklung - Methoden und Ergebnisse. - 210 S., Düsseldorf (Abt. Geologie der Universität).

SCHIRMER, W., 1981: Abflussverhalten des Mains im Jungquartär. - Sonderveröff. Geol.Inst.Univ.Köln, 41: 197-208, Köln.

SCHIRMER, W., 1983a: Criteria for the differentiation of the Late Quaternary river terraces. - Quaternary Studies in Poland, 4: 199-205, Warszawa, Poznan.

SCHIRMER, W., 1983b: Die Talentwicklung an Main und Regnitz seit dem Hochwürm. - Geol.Jb. A 71: 11-43, Hannover.

SCHIRMER, W., 1985: Landschaftlich-geologische Gliederung der Oberrheinebene nördlich Strassburg. - In: SCHIRMER, W. & K. STRIEDTER: 4-6.

SCHIRMER, W. & K. STRIEDTER, 1985: Alter und Bau der Rheinebene nördlich von Strassburg. - In: HEUBERGER, H. (Hrsg.): Exkursionsführer II: Unterelsass (Rheinebene N Strassburg), Lothringische Vogesen, 3-14, Hannover (Deutsche Quartärvereinigung).

TULLA, J.G., 1838: Carte über den Lauf des Rheins von Basel bis Lauterburg in 18 Blättern nach dem Zustand des Stromes vom Jahr 1838. - Nachdruck: Lehrwerkstätte für Flachdruck und Lithographie an der Kreisberufschule Waldkirch/Breisgau.

Lake, Mire and River Environments, Lang & Schlüchter (eds)
© 1988 Balkema, Rotterdam. ISBN 90 6191 849 9

Tectonic, anthropogenic and climatic factors in the history of the Vistula river valley downstream of Cracow

L.Starkel
Institute of Geography, Polish Academy of Sciences, Cracow, Poland

ABSTRACT: The area under investigation is located downstream of the Cracow Gate in the western corner of the Sandomierz Basin. The catchment area exceeds 8500 km^2. Vistula river gently meandering has a gradient 0.3-0.4 ‰ and in this reach is supplied mainly by small left-side tributaries from loess plateaus and not by the right-side Carpathian rivers.

Below the loess-covered pleniglacial terrace extends the 5-8 km wide Holocene alluvial plain 4-6 m high and narrow channel-side benches of 2-3 m high actual flood plain. On this Holocene plain we can distinguish few generations of the palaeochannels starting from straight streamways covered by Lateglacial deposits and Lateglacial palaeo-meanders below the edge of loess terrace filled by organic deposits or burried by sub-atlantic aggradation. Next generation was formed during the Atlantic phase, partly burried by Neolithic slope wash and in the eastern part well preserved due to late-Atlantic avulsion of the main channel. A specific feature of this reach is the aggra-dation of the Late Roman - mediaeval age. The remains of the oak forest cut by men in 1st to 2nd century AD is now covered by 5-6 m thick member of sandy bars and loamy over-bank deposits. The system of youngest palaeomeander formed during last millennium shows a tendency to increase of meander radius from 200 to 600 m. From the early of 19-th century starts a downcutting of river channel, decreasing downstream. In comparison to other parts of the Subcarpathian basins reach is characteristic by a lack of dissection of Lateglacial erosional level and increased aggradation during last two millenia. This fact can be related to the gentle subsidence in this part of the Sandomierz Basin.

1 INTRODUCTION

In the Lateglacial and Holocene history of valley floors we usually like to de-cipher a reflection of climatic changes. In the river valleys of Central Europe it is visible through the tendency to down-cutting and change from braided to meander-ing channel pattern (FALKOWSKI 1975, KOZARSKI and ROTNICKI 1977, SCHIRMER 1983, STARKEL 1983). But at the same time we find a lot of information on the aggra-dation in the subsiding basins (BORSY and FELEGIHAZY 1983), as well as in the areas deforested a long time ago (KLATKA 1958, MENSCHING 1950, HAVLICEK 1983, STARKEL et al. 1981). In some regions the role of non-climatic factors can be so strong that it is difficult to distinguish the impact of climatic variations. Particular authors try to reconstruct the tectonic movements

on the basis of the channel form and litho-logy (SHANCER 1951) or explain all fine-grained alluvia by accelerated soil erosion (MENSCHING 1950).

I will try to discuss this problem taking as an example the reach of the Vistula river valley in the western part of the Sandomierz Basin, in the Carpathian foreland, downstream of the narrowing in the Cracow Gate. In the upper part of the Vistula river basin we find a lot of examples of the reflection of climatic variations of the glacial-interglacial range as well as of a rhythmicity of a duration, ca. 2000-2500 years (RALSKA-JASIEWICZOWA and STARKEL 1975, STARKEL et al. 1981, SZUMANSKI 1982). The western part of the Sandomierz Basin shows, how-ever, a tendency to subsidence during the Quaternary, which is reflected in the extraordinary thickness (up to 50 m) of

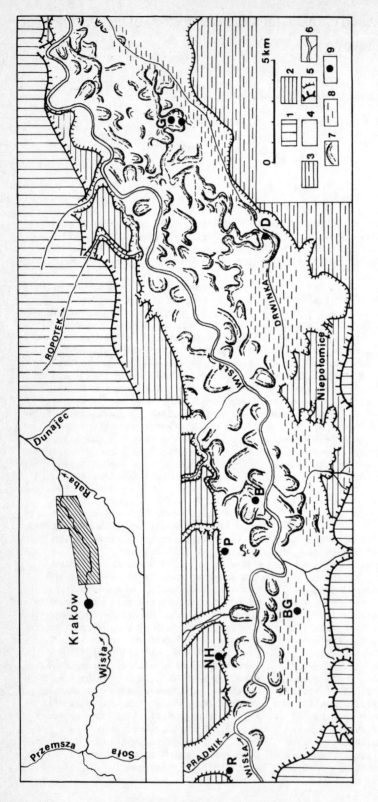

Figure 1. Relief of the Vistula river valley downstream of Cracow (RUTKOWSKI, STARKEL 1984). - 1 uplands, 2 - loess terrace, 3 - other Pleistocene sandy levels, 4 - Holocene alluvial plain, 5 - escarpments and terrace edges, 6 - Vistula river channel, 7 - paleochannels, 8 - wide marginal depressions (Lateglacial), 9 - localities dated by radiocarbon method, (R - Rondo, NH - Nowa Huta, P - Pleszów, B - Branice, Bg - Brzegi, D - Drwinka, G - Grobla forest).

Quaternary deposits below the valley floor near the outlet of the Raba river (POLTO-WICZ 1967), as well in the subsidence going on in the Oświęcim Basin (KOWALCZYK 1964, NIEDZIALKOWSKA et al. 1985). In the meantime, the tributaries of the Vistula in the western part of the Sandomierz Basin are draining the surrounding loess plateaux with a very old tradition of agriculture going back to the early Neolithic time (7000-6000 y BP). In such tributary valleys as Nidzica the aggradation going on during the best part of the Holocene is dominated by the suspended load, closely related to the rate of soil erosion (SNIESZKO 1985).

Therefore, I find this part of the 1090 km long Vistula river as a good example for discussing the overlapping of climatic, tectonic and anthropogenic factors in the last stage of the valley floor history.

2 CHARACTERISTICS OF FORMS AND DEPOSITS

The considered 35 km long reach is located between 185 and 220 km of the Vistula river and its catchment basin exceeds 8500 km². The Vistula river gradient varies between 0,2 and 0,4 ‰. Its sinuous and regulated channel amounts to 100 m in width. The water level rises usually after continuous rains, when the discharge exceeds 2'000 m^3 sec^{-1} (PUNZET 1981). The annual sediment load is higher than 340'000 t (BRANSKI 1975). Since the beginning of the XIXth century the downcutting of the river has continued and amounts up to 4 m in Cracow, and decreases to 1 m near the junction with the Raba river (TRAFAS 1975).

The valley bottom is bordered in the north by the loess terrace of Vistulian age rising 15-25 m above the main river (MAMKOWA, SRODON 1977) by the edge of the Proszowice Plateau and in the south by fragments of Pleistocene terraces and an alluvial fan of the Carpathian tributary, the Raba. The valley bottom is wider towards the east and occupied by several alluvial fills of the Holocene terrace 4-6 m high, after the regulation in the latter part of the XIXth century mainly overloaded (Fig. 1). At present, the Vistula channel is bordered by dams. The base of the alluvial fill is located at a depth 4-8 m below the water level in the western, and slopes to 20-50 m in the eastern part. The gravel member, differing

in age and thickness, is overlain with well-sorted sands deposited as the channel bars or levees. The topmost 1-3 m overbank loams are replaced - at some distance from the channel, especially in abandoned channels - by organic deposits, slope deluvia, and alluvial fans of the tributary creeks (STARKEL in: RUTKOWSKI and STARKEL 1984).

In the western part, in the transversal section there can be distinguished 6 zones. From the north, at the base of the loess terrace there is a very flat plain with completely filled paleochannels, indicated by stripes of wet meadows and arc-like edges of the loess terrace. Two generations of abandoned channels were distinguished there. To the west there are paleomeanders above 100 m wide sloping to the present river level. The erosional floors beneath indicate a phase of lateral migration and erosion. Their lateglacial age has been proved at the Rondo locality by younger Dryas muds and peat dated 9390±180 y BP (MAMAKOWA 1970) and near the Central Square in Nowa Huta by clays and peat with their bottom date 9660±110 y BP (KALICKI in: RUTKOWSKI and STARKEL 1984).

To the east the paleomeanders have been partly fossilised by deluvia. These channels date back to the Atlantic phase, but unlike the Wisloka valley (STARKEL et al. 1981) they are not deeper but shallower if compared with the Lateglacial. At Pleszów, there was found a close connection of filling with human activity. Below the horizon dated 6255±40 y BP the first cultivation phase was found. The soil erosion ca. 6000 y BP was so intensive that delivial loams started to overlie swampy oxbow lake deposits (WASYLIKOWA et al. 1985). An increase in humidity and rise in the ground water level followed between 5480 and 4750 y BP - which correlates well with the widespread humid phase (RALSKA-JASIEWICZOWA and STARKEL 1986). The channel itself was probably left by the Vistula during the older phase of an increased fluvial activity 6500-6000 y BP (STARKEL 1983).

The next zone of the transversal profile is characterised by very sinuous meanders up to 50 m wide with poorly developed pointbars. The wood found in the sandy fill was dated 5190±70 y BP. This system was probably left by the Vistula at the break of the Atlantic and Subboreal periods.

The zone closer to the present-day channel is up to 1.5 km wide and dissected by

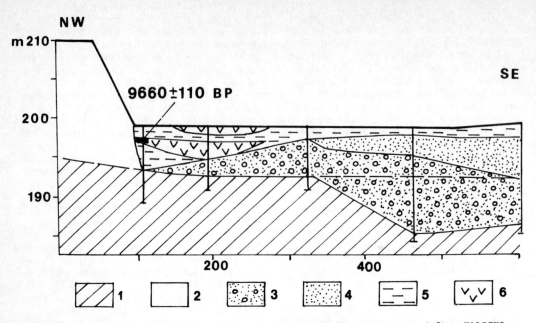

Figure 2. Lateglacial erosional surface and paleochannel at Nowa Huta (after KALICKI, in: RUTKOWSKI, STARKEL 1984). 1 - Miocene clays, 2 - deposits of loess terrace, 3 - gravel and sand, 4 - sand, 5 - alluvial loam, 6 - peat.

young paleomeanders with well-preserved point bars (radius > 350 m, width > 100 m). The youngest were active up to the XVIIIth century. In the gravel pit in Branice-Stryjów, fossilised paleochannels were exposed with bottoms sloping 1-2 m below the present-day Vistula river. The oldest of them is from the Alleröd (dating of the fill 10920$\pm$230 y BP), the two younger fills are dated 2200$\pm$70 and 1480$\pm$260 y BP (Fig. 4). This indicates that the Vistula river reached several times a similar niveau and was not strong enough to cut a trench in the coarse gravel. But at a similar level, 6 meters below the surface, there were found in sands many stumps growing on the existing flood plain and cut by using axes in the I-II century A.D. Just at that time, on the nearby loess terrace there was florishing pottery and iron smelting. The deforestation of the flood plain accelerated the river channel's mobility. And in fact the aggrading Vistula rose its channel niveau, which is indicated by the gravel lenses in the sandy member 2-3 meters higher. The overlying upper sands belong to the system of greater meanders active in the last millenium. This aggradation phase is also known

from the Cracow centre where the cultural layers were buried by alluvial loams in the second half of the XIth and in the XIIIth century (RADWANSKI 1972). Finally, in the XIX-XXth century the river channel was deepened by ca. 3 meters.

At the right bank of the Vistula, outside the dams, there is visible a similar belt with bigger meanders, and next a 2 km wide flat and swampy depression with rare remains of buried paleochannels. In the gravel-pit Brzegi there were excavated channel facies gravels and sandy bars with plant detritus (Pinus, Betula). The trunk buried in gravels dated at 10690$\pm$190 y BP place the accumulation phase in the Younger Dryas. This aggradation reached 2-3 m above the present mean level of the Vistula.

In the eastern part the differentiation of the valley bottom has another character (Fig. 1), because the Vistula left its central position and undercut the edge of the loess terrace as well as of the Proszowice Plateau (STARKEL 1967, STARKEL, GEBICA in: RUTKOWSKI and STARKEL 1984).

Starting from the north, the zone 1.5 km wide forms an active plain with river, embankments and paleomeanders of a larger

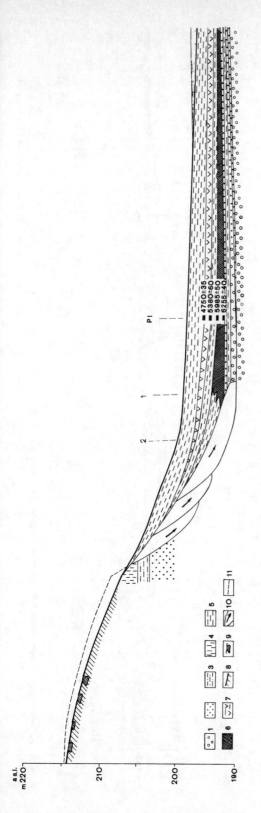

Figure 3. Edge of loess terrace and fossilised paleochannel at Pleszów (after WASYLIKOWA et al. 1985). 1 – gravels, 2 – sands, 3 – sandy muds, 4 – loess, 5 – deluvial loams, 6 – peat, 7 – peaty muds, 8 – man-made pits, 9 – fossil landslides, 10 – primary slope profile.

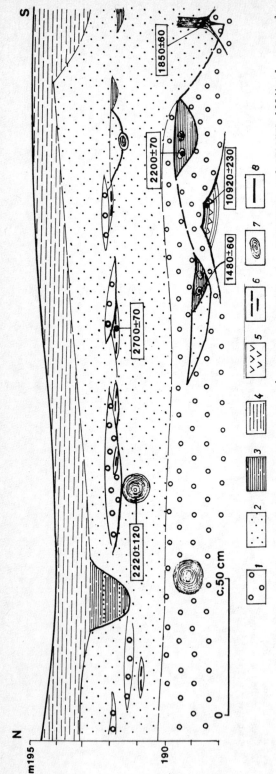

Figure 4. Simplified section across gravel-pit in Branice-Stryjów (after STARKEL; in: RUTKOWSKI and STARKEL 1984). 1 – gravels with sand, 2 – sands, 3 – clays and silty muds, 4 – silty and sandy alluvial loams, 5 – peaty muds, 6 – lenses of organic detritus, 7 – trunks, 8 – erosional surface. Note: upper trunks are redeposited (radiocarbon datings older than below).

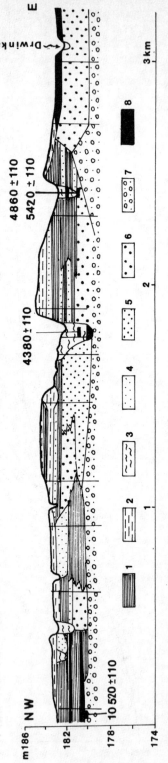

Figure 5. Section across the abandoned paleochannel system in the Grobla forest (after BZOWSKI 1973, radiocarbon datings added). 1 – clays, 2 – loams, 3 – swampy deposits (paleochannel fills), 4 – loamy and fine sands, 5 – medium-size sands, 6 – coarse sands, 7 – gravels and sands, 8 – organic deposits.

166

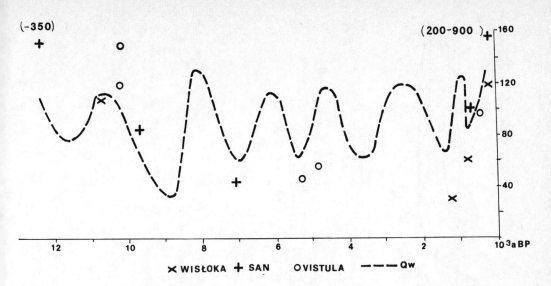

Figure 6. Change of paleochannel width in time in the Vistula valley reach compared with its tributaries: Wisloka (STARKEL et al. 1981) and San (SZUMANSKI 1982) Qw - variations in flood frequency (after STARKEL 1983).

size (r = 250-400 m, w = 60-80 m). Some of them were abandoned in the XVIIIth-XIXth centuries. Above the plain, two hills are rising 15 and 13 m high. The Trawniki hill, above 300 m long, is built of sands with muds modified by load structures indicating an active layer of the permafrost. They are overlain with 8 m thick loess member with snails, characteristic of the shallow water bodies of an alas type (after preliminary information by S.W. ALEXANDROWICZ). The other hill, Skala, exposed pantry hollows on its gentle slope of late Roman and early - mediaeval age, overlain with alluvial loams. This indicates a much lower position of the river channels, and very rare floods at that time (ZAKI et al. 1970). The lithology of both hills is identical with the structure of the left-bank loess terrace of the Vistulian age. This can be only explained as the separation of two hills from the wide loess terrace by the meandering river. The facts, described below, show that this happened in the Neoholocene (Fig. 5).

To the south, there occur smaller paleomeanders (r = 150-220 m, w = 40-55 m) entering the wide levee 1-2 m higher, which is cut across by some crevasses. Towards the east it changes into a wet plain built of peaty clays, described by

BZOWSKI (1973) as lateglacial deposits. Now, dating from a depth of 2.1 m (10520 ± 110 y BP) supported this view.

To the south there extends a slightly elevated zone rich in paleomeanders composed of 2 systems (BZOWSKI 1973): The older one includes separate abandoned channels (r = 120-235, w = 45-55 m). The channels of the younger system 45-70 m wide have a greatly changing curvature (from sinuous to meandering with r = 125-165 m). These channel parameters and the narrow levees suggest that the Vistula started at first to adapt its channel to more frequent floods and then there followed a total avulsion of the channel to the north. In the older system, the bottom of the fill was dated at 5420 ± 140 y BP, and its middle part at 4860 ± 110 y BP. The dating of the middle part of the younger system fill is 4380 ± 110 y BP. These data indicate that the Vistula started to straighten its channel before 5400 y BP, and the avulsion followed probably at the beginn of the Subboreal together with a rise in flood frequency.

The southern part of the valley floor 1-2 km wide is occupied by a flat and wet depression drained by the Drwinka stream. Built of sand with a silty cover it retains a character of a braided river from

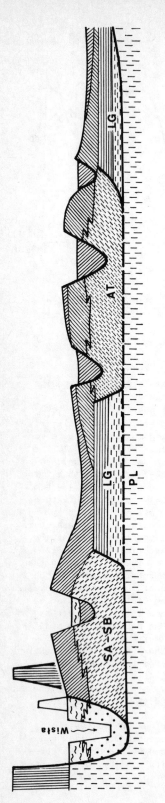

Figure 7. Scheme of the Holocene aggradation in the western part of the Sandomierz Basin (with subsidence tendency). 1 – channel facies (Pleniglacial), 2 – loess (PL), 3 – channel facies (Lateglacial), 4 – alluvial loams (LG), 5 – channel facies (Atlantic), 6 – alluvial loams (AT), 7 – channel facies (Subboreal-Subatlantic), 8 – alluvial loams (SB–SA), 9 – channel facies (19–20th century), 10 – alluvial loams (19–20th century), 11 – erosional contacts, 12 – dams.

the Lateglacial or even older phase. At
the margin of this zone, in the poorly expressed big paleomeander ca. 150 m wide,
P. GEBICA found below the flood loam a
peat layer and gythia. The radiocarbon
date from the peat base 8760±90 y BP indicates that this channel was active probably earlier (before the Holocene). The
age of the top of the peaty layer (7980±80
y BP) covered with overbank loams can be
explained as the beginn of flooding in
the early Atlantic time, well known from
the Wisloka valley (STARKEL et al. 1981).

3 CONCLUSIONS

The evolution of the Vistula valley
reaches here described, lying in the
western part of the Sandomierz Basin
during the Lateglacial and Holocene can be
explained by different climatic, tectonic
and anthropogenic factors overlapping one
another.

The climatic factor is reflected in the
change of channel parameters well known
from other valleys in the upper Vistula
catchment basin (Fig. 6) starting from
the large Lateglacial paleochannels through
smaller ones of the Holocene age, up to
large ones formed during the last centuries
(cf. SZUMANSKI 1982, STARKEL et al. 1981).
The channel avulsions and upbuilding of
peat by alluvial łoams also followed
during the phases with a higher flood
intensity described from many other Central
European valleys (Fig. 7). But it is very
difficult to explain the absence of the
vertical downcutting during the Holocene
in another way than by the subsidence
tendency, which in the eastern part continues from the Pleistocene (POLTOWICZ
1967). Therefore we observe channel
avulsions, so characteristic of the subsiding basins (Hungarian Basin, Hindustan
Basin - see STARKEL 1972). On the con-
trary, the present-day channel downcutting
is so gentle (TRAFAS 1975). And even the
slope of the Holocene floodplain decreases
from 0.5 %o in the west to 0.3 %o in the
east.

This change in the floodplain gradient
can also be explained as the result of
accelerated aggradation of the Vistula
river and its tributaries at the outlets
from narrow reaches. These fans are built
of the product of soil erosion, starting
in the Neolithic (WASYLIKOWA et al. 1985),
especially intensive during the Late-Roman
and Mediaeval periods. The channel down-

cutting in the last centuries is also
caused by man, but it decreases in the
subsiding area. The reach discussed is a
good example for the overlapping of different factors, and shows how careful we
must be, when drawing conclusions of the
paleogeographic characters.

4 ACKNOWLEDGEMENTS

The Author wishes to express his cordial
thanks to T. KALICKI and P. GEBICA for
close cooperation in field work and elaboration of results. A joint paper is under
preparation. Great help was given by
K. WASYLIKOWA, D. NALEPKA and D. TOMCZYNS-
KA in the preliminary identification of
the pollen spectra and macrofossils, as
well as by M. PAZDUR in radiocarbon dating
of the organic horizons and buried oaks.

5 REFERENCES

BORSZY, Z. and FELEGYHAZI, E., 1983,
Evolution of the network of water courses
in the North-Eastern part of the Great
Hungarian Plain from the end of the
Pleistocene to our days.- Quaternary
Studies in Poland, 4, Poznań, 115-124.
BRANSKI, I., 1975, Ocena denudacji dor-
zecza Wisly na podstawie wyników pomia-
rów rumowiska unoszonego. Prace IMGW, 6,
5-58.
BZOWSKI, M., 1973, Rzeźba i stosunki wodne
dna doliny Wisly w rejonie północnej
części Puszczy Niepolomickiej. Zakl.
Ochrony Przyrody PAN, Studia Naturae,
ser. A, 7, 7-37.
FALKOWSKI, E., 1975, Variability of channel processes of lowland rivers in
Poland and changes of the valley floors
during the Holocene. Biul. Geolog. U.W.,
19, Warszawa, 45-78.
HAVLICEK, P., 1983, Late Pleistocene and
Holocene fluvial deposits of the Morava
River (Czechoslovakia). Geolog. Jahrbuch, A71, 209-217.
KLATKA, T., 1958, Muly antropogeniczne
doliny Swiśliny i ich dynamiczna inter-
pretacja. Acta Geogr. Univ. Lodziensis,
8, 165-183.
KOWALCZYK, Z., 1964, Analiza wyników badań
geodezyjnych nad wspólczesnymi natural-
nymi ruchami powierzchni pld. części
Górnego Slaska Prace Kom. N. Techn. Oddz.
PAN, Kraków, Geodezja 1, 61 p.
KOZARSKI, S. and ROTNICKI, K., 1977, Valley
floors and changes of river channel

pattern in the North Polish Plain during the Late-Würm and Holocene. Quaestiones Geographicae 4, Poznań, 51-93.

MAMAKOWA, K., 1970, Late Glacial and Early Holocene vegetation from the territory of Cracow. Acta Palaeobot. 11, 1, Kraków.

MAMAKOWA, K., SRODON, A., 1977, O pleniglacjalnej florze z Nowej Huty i osadach czwartorzedu doliny Wisly pod Krakowem. Rocz. Pol. Tow. Geol. 47, 4, 485-511, Kraków.

MENSCHING, H., 1950, Schotterfluren und Talauen im Niedersächsischen Bergland, Gött. Geogr. Abhandl. H. 4, 51 p.

NIEDZIALKOWSKA, Ewa, GILOT, E., PAZDUR, M. and SZCZEPABEK, K., 1985, The Upper Vistula valley near Drogomyśl in the Late Vistulian and Holocene. Folia Quaternaria, Kraków, in print.

PUNZET, J., 1981, Zmiany w przebiegu stanów wody w dorzeczu górnej Wisly na przestrzeni 100 lat (1971-1970). Folia Geographica, 14, 5-28.

RADWANSKI, K., 1972, Stosunki wodne wczesnośredniowiecznego Okolu w Krakowie, ich wplyw na topografie osadnictwa próby powiazania tych zjawisk ze zmianami klimatycznymi. Mat. Archeologiczne, 13, 5-40.

RALSKA-JASIEWICZOWA, M. and STARKEL, L., 1975, The basic problems of palaeogeography of the Holocene in the Polish Carpathians. Biul. Geol. UW, 19, Warsawa, 27-44.

RALSKA-JASIEWICZOWA, M. and STARKEL, L., 1986, Record of the hydrological changes during the Holocene in the lake, mire and fluvial deposits of Poland (in print in Folia Quaternaria).

RUTKOWSKI, J. and STARKEL, L. (eds.), 1984, Holocen okolic Krakowa (materialy Sympozjum), Wyd. AGH, 95 p.

SCHIRMER, W., 1983, Criteria for the differentiation of late Quaternary river terrace. Quaternary Studies in Poland 4, 199-205.

SHANCER, E.V., 1951, Alluvia of the lowlands of the temperate zone and their role for knowledge of the regularities of alluvial beds formation (in Russian). Trudy Geol. Inst. Akademii Nauk SSSR, 135, Moscow.

STARKEL, L., 1967, Wisla wśród gór wyzyn. Przewodnik geol. z biegiem Wisly cz. I, Geol. 31-159.

STARKEL, L., 1972, Trends of development of valley floors of mountain areas and submontane depressions in the Holocene. Studia Geom. Carp. Balc. 6, 121-133.

STARKEL, L., 1983, The reflection of hydrologic changes in the fluvial environment of the temperate zone during the last 15'000 years; in: Background to Palaeohydrology: A Perspective, ed. K.J. GREGORY, J. Wiley 2/3, 235 p.

STARKEL, L. (ed.), S.W. ALEXANDROWICZ, K. KLIMEK, A. KOWALKOWSKI, K. MAMAKOWA, E. NIEDZIALKOWSKA, M. PAZDUR, L. STARKEL, 1981, The evolution of the Wisloka valley near Debica during the Lateglacial and Holocene, Folia Quaternaria, 53, Kraków, 91 pp.

SZUMANSKI, A., 1982, The evolution of the Lower San river valley during the Lateglacial and Holocene. Prace Geograficzne IGiPZPAN, spec. issue 1, Warsaw, 57-78.

SNIESZKO, Z., 1983, Wyksztalcenie osadów póznovistuliańskich i holoceńskich w rejonie Dzialoszyc. Przewodnik konferencji nt. Póznovistuliańskie i holoceńskie zmiany środowiska geograficznego na obszarach lessowych Wyzyny Miechowskiej i Opatowsko-Sandomierskiej Uniw. Slaski, Katowice, 38-49.

TRAFAS, K., 1975, Zmiany biegu koryta Wisly - na wschód od Krakowa w świetle map archiwalnych i fotointerpretacji. Prace Geogr. IGUJ, Kraków, 40, 85 p.

WASYLIKOWA, K., STARKEL L., NIEDZIALKOWSKA E., SKIBA S. and STWORZEWICZ E., 1985, Environmental changes in the Vistula valley at Pleszów caused by Neolithic man. Przeglad Archeologiczny (in print).

ZAKI, A., FRAS, M. and OLSZOWSKI J., 1970, Stanowisko wczesnośredniowieczne w Grobli, pow. Bochnia. Spraw. z Pos. Kom. Nauk. Oddzialu PAN, Kraków, 14, 486 p.

Lake, Mire and River Environments, Lang & Schlüchter (eds)
© 1988 Balkema, Rotterdam. ISBN 90 6191 849 9

The Pleistocene-Holocene boundary in the Po valley south of Torino (NW Italy)

D.Tropeano
CNR, IRPI, Torino, Italy

ABSTRACT: The sequence of Quaternary deposits in the subsoil of the Po River alluvial plain to some 30 km South of Torino has been studied by borehole data and gravel pit observations. The data collected up to now suggest that a stratigraphic gap may exist between the end of the Würm and the base of the Holocene sediments and that the peat horizon, where present, may be considered as a reference level to define the overlying sediments as being of Holocene age.

1 INTRODUCTION

At C.N.R.-Istituto di Ricerca per la Protezione Idrogeologica nel Bacino Padano, research is carried out on stream activity in Northern Italy, special attention being paid to river channel changes with regard both to planimetry and vertical profile. Experimental studies on erosion and sediment transport are in progress as well. Such research is carried out also with a view to apply the results to the problems relating to geological hazard evaluation. The aim of this report is to give a merely cognitive contribution to a better knowledge of the Quaternary of the Po Valley, that is, to a better understanding of the fluvial processes which occurred in the past along the largest Italian river.

Observations made in the last two years in ten active gravel pits along a reach of the Po River of some 30 km South of Torino, were integrated by stratigraphic data supplied by over 300 drillings, more or less homogeneously distributed over an area of about 290 km². This area, forming part of the Western Po Valley, has a flattened surface and is irregularly defined on the right by the Torino hills and the "Poirino Plateau", while it opens on the left to the high plain forming the Alpine piedmont. In this area, deposits belonging to the following chronostratigraphical sequence have been generally recognized:

1) Pleistocene (sensu lato). A complex of sediments up to several tens of metres thick, for which it is quite impossible, for the time being, to give a more detailed definition. In this complex the upper layers belong at least in places to early and middle ages of Würm.

2) Würm 3 stage. It is mainly represented by 5 to 30 m thick backswamp deposits, alternating to lenses of gravel, dating back to about 24'000 to 30'000 radiocarbon years BP.

3) Holocene (middle to recent). There is no evidence up to the present of deposits dating back before the Atlantic period. The whole thickness of Holocene sediments seems not to exceed 10-12 m.

2 LITHOFACIES DESCRIPTION

With the only exception of deposits belonging to the Holocene, which are in most cases to be seen directly on the walls of gravel pits and on the stream banks, for a depth of 5-7 m above the water table, the only exception being, in few cases, the top of the sandy-pelitic complex ascribed to Würm 3. Very rarely, during gravel excavations, was groundwater artificially lowered down to a depth of 12-15 m thus permitting more extensive observations.

Information about grain-size, lithology, colour, texture of the sediments was necessarily drawn by observations on materials dredged up from the quarries, supplemented by an almost approximate indication given by quarrymen's experience about the different depths where the excavated material had

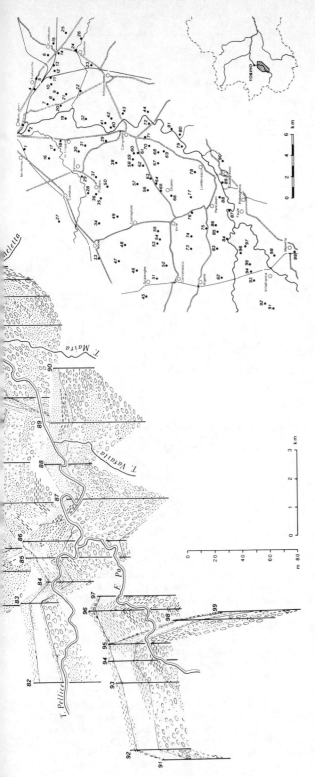

Figure 1. Fenced diagram showing lithostratigraphic characters of the investigated area. The following units can be identified: I, II – Pre-Quaternary deposits (I – marine Pliocene marls and fine sands; II – "Villafranchian" shales); 1 – Pleistocene (Würm p.p.): yellowish coarse gravels (1a), sands (1b), pelite (1c); 2 – Upper Pleistocene (mainly Würm 3): greyish gravels (2a), sands (2b), sandy-clayey silts (2c); 3 – Holocene: gravels (3a), sands (3b), silts (3c). Some of the stratigraphic columns refer to two drillings or more. The diagram thus actually represents a total of 154 logs. In 15 % of cases full sequences of core samples were obtained.

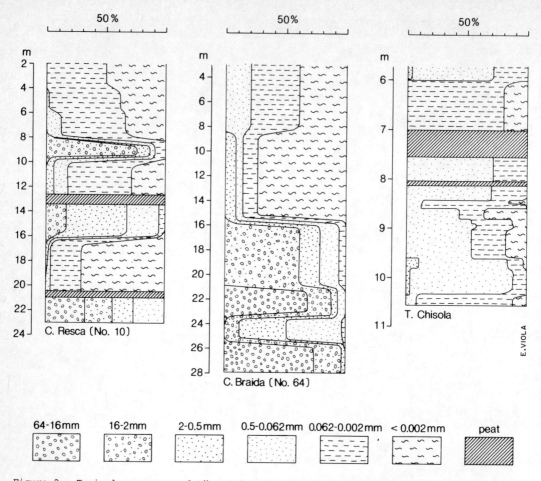

Figure 2. Typical sequences of Würm 3 deposits in the area under study. Changes in grain-size distribution may occur both on a large and small scale. The first two columns are details of the same data shown in Fig. 1.

been found. Most useful were also the stratigraphic columns drafted by drilling enterprises, on condition that the data were directly verified when possible. Sediment core samples were collected for grain-size and petrographic analyses, taking advantage of a series of drillings quite uniformly scattered in the area under study, for a total of 88 samples. The stratigraphic columns, properly interpreted, were interpolated in the form of a fenced diagram, depicted in Fig. 1.

The sediments ascribed, as a whole, to Pleistocene s.l. are mainly gravel-sand mixtures in loamy-sandy matrix. Gravel content varies widely from 25 to 90 per cent

of the sample. Median diameter is comprised between 0.42 and 20 mm. Cobbles and small boulders are locally present with a maximum length of up to 350-400 mm. A horizon characterized by the occurrence of large-sized individuals can be found at intervals at a quite uniform depth 25-30 m below ground surface, over a long reach, from Moncalieri to Faule. Nearly always a more or less oxidized coat characterizes the coarser fraction while the colour of the matrix may vary quite independently of the depth, from mostly yellowish to yellowish-grey, grey, brown and more rarely reddish. Intercalations of finer sediments are locally present; in particular, a

Figure 3. Gravel pit near Carignano (Monticone Quarry), where the groundwater table was artificially lowered (Sept. 1984). The following litho-
facies, all belonging to Würm 3, were thus visible: peat layer (A); sandy
silt loam, peat rich in frustules (B); blue-greyish sands (C). The boundary
with gravelly-sandy Holocene deposits is apparent and marked by an erosional
surface (dashed line). Scour grooves are well evidenced on the collapsed
block in the right (photo processing P.G. TREBO', C.N.R.-I.R.P.I., Torino).

Figure 4. Typical Holocene deposits, left
of the Po River, near La Loggia. The lower
two-thirds of this section are represented
by sand, gravel and cobbles, often incor-
porating brick fragments, overtopped by
grey-yellowish silts and fine sands, corresponding to present-day overbank deposits.

sandy-pelitic complex, yellowish to blue-
grey in colour including peat layers
occurs in several places at a depth roughly between 40 and 50 m from the ground surface. Palynologic analyses being made on
such deposits suggest preliminarily that
they may be of Lower-Middle Pleistocene
age (Dr. E. CERCHIO, personal communication).

In the area considered here alluvial
Quaternary deposits overlie with a marked
erosional contact the sediments of various
origin of the top of the Neogene of the
Torino hill anticline. This structure is
partly buried under the Po plain, its axis
dipping South-Westward. Pleistocene deposits are thus directly superimposed at
first on grey sands, sandstones and marl,
belonging to marine Pliocene, often with
the interrelation of a conglomerate including remanied Pliocene Mollusca (TROPEANO
et al. 1984). Furthermore, the substratum
is represented by prevailing blue-greyish
shales, with lenses of small-sized gravels,
which can be considered as "Villafranchian"
deposits. Due to the progressive dipping of
these formations from North to South, the
thickness of Pleistocene deposits increases
rapidly, from a few metres near Moncalieri

Figure 5. Subfossil oak tree, found with many others in a gravel pit opened in 1970 right of T. Banna inflow. Similar buried log deposits can be found throughout the floodplain of Po River in all the area here considered.

up to over 70 m in the Carignano-Carmagnola area. Although it is very difficult, on the basis of the present data, to chronological-ly differentiate all the deposits called "Pleistocene" as a whole, it is reasonable to think their upper part as belonging to Würm. At Molinello Quarry (Moncalieri) wood fragments found in coarse gravel deposits directly laid on Pliocene at a depth of 20-25 m from the soil surface were radiocarbon dated $\geq$ 44'000 y BP. The measurements were made by Laboratorio per le datazioni con il Carbonio-14 dell' Università degli Studi di Roma "La Sapienza" and the data are supplied by Prof. C. CORTESI, which is kindly acknowledged. Only a few metres above, remains were found of Mammuthus primigenius and Megaloceros giganteus, as reported by CHARRIER & PERETTI (1977). It can thus be assumed that Würm extends throughout the whole section of the gravel pit, even if the lower chronological limit is unknown.

The first ^{14}C datable layers overlaid on the above-described deposits form part of a widely spread sequence of sandy-pelitic sediments, blue-greyish in colour, sometimes alternating with subordinate gravels. One or two main peaty layers, usually 25-30 cm thick, are often associated with the pelitic facies. Such bog deposits were already described by CHARRIER & PERETTI (1975, 1977), as occurring in four sites in the Cadiolo-Vinovo-La Loggia-Moncalieri areas, respectively. Based on new strati-graphical data, this horizon appears much more widespread, extending also Southward; his planimetric extent roughly corresponds to the area depicted in Fig. 1. An average thickness of around 15-20 m can be attributed to such complex, from statistical examination of boring logs. Considering the vertical distribution of the sediments, on the whole 49 % of these are silty clay loam (KAROL 1960), 43 % sands, 8 % medium to fine prevailing gravel, sometimes in hard-cemented levels. These sediments are of course randomly distributed, as is the rule for all fluvial deposits, but locally, also on a small scale, full sequences from finer to coarser and vice-versa can be observed (Fig. 2). Peat intercalations may be present at various levels. One of these, presumably at the top of the series near Candiolo, was radiocarbon dated at 24'300±200 y BP; another one in the vicinity, near Vinovo, revealed an age of 26'270±400 y BP (CHARRIER & PERETTI 1977). The radiometric

176

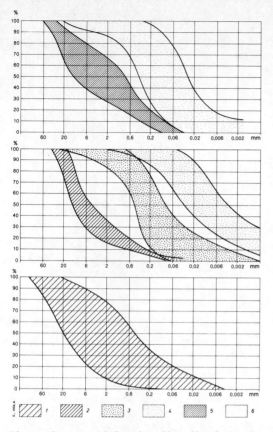

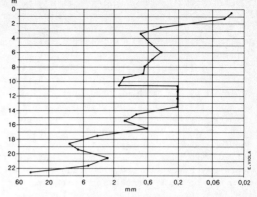

Figure 6b. Grain-size variations depth-wise, expressed by D_{50} values, near the confluence of the Po-Chisola streams (site 2 in Fig. 1). The graph is based on grain-size analyses, made on a total of 58 core samples taken at 1 m distances. Although some fluctuations occur, the largest of which, at a depth of between 10.5 and 13.5 m, is due to the presence of the sandy-pelitic complex, a progressive downward increase in size is evidenced. The maximum value is attained at a depth of 23 m, at the boundary with Pliocene. A similar trend can be observed in various other stratigraphic columns. (Drawn by elaboration of data kindly supplied by "Azienda Energetica Municipiale", Torino).

Figure 6a. Particle-size distributions of Quaternary alluvial deposits in the Po Valley upstream of Torino. 1 - Pleistocene gravels, at a depth of more than 24-26 m (36 samples); 2, 3, 4 - Würm 3 deposits (gravels, sands and pelite, respectively; 29 samples); 5 - Holocene gravels (Middle to Recent; 15 samples); 6 - present-day overbank deposits (8 samples). Grain-size analyses made by R. MASSOBRIO, C.N.R.-I.R.P.I., Torino.

age of a wood included in peat, some 8 km Eastward, near Trofarello, was 30'660±1290 y BP. By contrast, a wood sample included at what seems to be the top of the series, close to the specimen of Vinovo, showed a ^{14}C age of ≥ 44'000 y BP. It is therefore reasonable to deem such stump remanied from older deposits, which are not yet identifiable in that area. Palynologic studies (CHARRIER & PERETTI 1977; TROPEANO & CERCHIO 1984) on peat samples collected at different sites evidenced a Pleniglacial

forest growth, with Pinus silvestris prevailing, occurrence of Betula nana and Artemisio and thermophile plants virtually lacking.

The upper part of the sandy-pelitic series described above outcrops only in few places, along the beds of Po and Chisola streams and, from time to time, in the gravel pits, left of the Po River, between Carignano and La Loggia. It will be seen that the contact with overlying deposits (all of these belonging to Holocene) is erosive throughout the above areas; by logical extrapolation it may be inferred that the Würm deposits in the section corresponding to the present-day alluvial plain of Po River end sharply with a stratigraphical gap. Fairly numerous borehole data reveal the presence of the sandy-pelitic series also in the subsoil of the left terrace of the Po River, between La Loggia and Faule. As already noticed by CHARRIER &

177

PERETTI (1977), this implies that sediments in that area, formerly mapped as Riss deposits in the 2nd Edition of "Carta Geol. d' Italia"(F^{O}. 68) belong at least locally to Würm. Just North of Carignano the terrace is cut by river bank erosion, showing from the bottom a grey pelite deposit with peat, followed by a 1 m thick conglomerate layer. Remains of Mammuthus primigenius were reported by PORTIS (1898) as having been found near La Loggia in an hard-cemented conglomerate, which may be reasonably identified with this level. Although rearrangements may have occurred, this suggests that Würm may be still present above the top of the sandy-pelitic series. For a few metres above the conglomerate, yellow sands with lenses of fine gravel, overtopped in turn by silt and a weathered horizon, are to be found; the whole of such deposits, some 8 m thick, should represent a later stage of Würm, although scarce rolled brick fragments found in place by the writer along the same terrace scarp, at a depth of 5 to 6 m below the topsoil, suggest that the Holocene presence cannot be excluded, at least locally.

Holocene sediments can be detected by the following criteria:
- radiocarbon dates of subfossil woods contained at given levels;
- evidence of artifact remains of historical age;
- much fresh appearance of the deposits (incohesive and unweathered);
- geomorphological evidence of present-day depositional processes, in some cases confirmed by historical documents, or by direct observations;
- correlation with other deposits, the Holocene age of which can be safely demonstrated.

The Pleistocene-Holocene boundary is well defined over a large part of the present alluvial plain, as said above, where the top layer of the sandy-pelitic series, mostly in facies of peat, occur. In such cases, the typical Holocene stratigraphy is expressed by two kinds of deposits:
- from the bottom, fine to coarse gravel, in a grey sandy matrix, ranging from about 30 to 65 per cent (D$_{50}$ from 0.7 to 12 mm). These deposits are divided from the underlying peaty-pelitic facies; still belonging to Würm 3, by a marked erosional surface; this is evidenced also by remanied blocks of peat, even 1 m large, which are often incorporated in the gravel, near the bottom. At this lev-

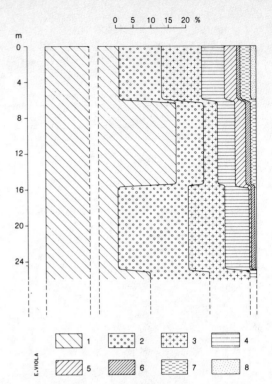

Figure 7. Lithologic content of the gravel samples examined (accounting for a total of 2366 individuals) belonging to Holocene (0-6 m), Würm 3 (6-16 m) and Pleistocene (below 16 m). Legend: 1 - Quartz, 2 - Permian conglomerate, 3 - Granite, 4 - Gneiss, 5 - Serpentine, 6 - Ophiolites l.s., 7 - Calcschists, 8 - Other rocks. Depth values are averaged from boring-log data; the occurrence of quartz is excluded from the percent scale, for plotting reasons. A heavy increase in Permian conglomerate pebbles can be seen below 16 m, as well as a progressive reduction in ophiolites and calcschistes from the surface down to around 25 m. Below such depth, quartz, conglomerate and granite represent nearly the whole content of the samples analyzed.

el, nearly coinciding with the water-bearing stratum, hardened crusts, a few cm thick, may occur. Sometimes pebbles are black or reddish coated by oxidation. Brick and tiles, and other artifact remains, dating back from Roman times to modern age, are widespread over a long reach, from Carignano to Moncalieri, and generally occur in the middle-upper section of the gravels;

178

- superimposed on such gravel deposits, which are in average 3-5 m thick, either medium to fine grey sands are to be found, with small percentages of gravel, or grey-yellowish silt, in which small amounts of fine sand and clay are also present (Fig. 4). Most of the silty-clayey sediments to the right of Po, South of Moncalieri, are to be intended as remanied by local streams, from the loessial deposits, more or less weathered by pedogenesis, partly covering the Southern slope of the Torino hills and the Poirino Plateau. This explains the dark yellowish or orange-brown colour of these fluvial sediments, the very recent age of which can be revealed near the Banna stream, by brick remains at 2.5-3 m below the ground, and by old chronicles reporting overflow depositions until the end of 18th Century (ANSELMO & TROPEANO 1978).

Based on the data collected up to now, the whole depth of Holocene deposits in the area under study averages 6-8 m, ranging from a minimum of 4-5 m (Moncalieri-La Loggia area) up to presumably 12-15 m, judging from the changes in colour and texture observed in several cores sampled at around this depth.

Radiocarbon dating of remanied trees, distributed in plenty in the reach of the Po Valley here considered (Fig. 5), should improve the present knowledge on accumulation times of the deposits during the Holocene in that area. A first assay was made in two sites: the first one at "Fontane" Quarry (Faule), where, at a depth of approx. 8 m the gravels contain a layer rich in remanied tree trunks, even of large dimensions (up to 15 m long); one of these showed an age of 5'855±75 y BP (Atlantic period). The other site is some 25 km North-Eastward, in the Rioverde plain, near Pralormo, where a gravel pit was opened in the past, revealing a big accumulation of buried logs at a depth of between 6 and 11 m. A wood specimen found by recent drilling in the same site, ^{14}C dated at 4'257±68 y BP, permits to locate the bottom of Holocene gravels (Subboreal) at a depth of about 10.5 m. Here Holocene gravels lie with apparent unconformity on Middle-Upper Pleistocene clay loams.

3 AN OUTLINE OF RIVER CHANNEL CHANGES DURING QUATERNARY IN THE AREA UNDER STUDY

Hypothesis on palaeohydrology of the Po Valley South of Torino have been made since the last Century up to recent times (GABERT, 1962; CARRARO, 1976); in particular, studies on the geomorphological evolution of the plain during Holocene are due to MARAGA (1983). The stratigraphical data acquired up to now make it possible to put forward a few remarks concerning sediment transport energies, which must have varied considerably in the past in the reach here considered.

With reference to the envelopes of grain-size curves reported in Fig. 6a, one can see that there is a general, progressive trend to the increase in the size of deposits, according to the depth, starting from the ground surface. For clarity, the grain-size curves have been grouped according to the lithostratigraphic units recognized. During part of Pleistocene (presumably before 30'000 y BP, or little more) braided patterns seem to have been prevailing, in connection with even conspicuous bedload discharges and able to carry large-size materials. Meandering episodes must have occurred as well, as documented by the fine-grained, peaty deposits found near Carignano and Carmagnola, at a depth of between 40 and 50 m.

According to the authors (GASTALDI, 1872; PREVER, 1907; SACCO, 1917; PEOLA, 1942; GABERT, 1962; PETRUCCI & TAGLIAVINI, 1968; CARRARO et al., 1969; BORTOLAMI et al., 1976), during Middle or Late Quaternary, at a time not yet accurately identified, River Tanaro, before his well-known flow diversion, joined the present Po plain in an area roughly comprised between Lombriasco and Carignano. Its passage is witnessed by pebbles of Permian conglomerate typical of the Upper Mountain Basin of Tanaro, and not present in the Alpine Po Valley. Lithologic analysis of gravel samples drawn from boreholes shows in some cases (e.g. near Osasio, Carignano, Trofarello) an abrupt increase in the percentage of pebbles derived from such conglomerate, around the average depth of approx. 16 m (Fig. 7). On this basis, one can reasonably express the hypothesis that deposits below such depth may be closely related to the palaeochannel of River Tanaro, while the overlying sediments, belonging to Würm 3, may originate in part from rearrangements of the above deposits, judging from the reduced percentage of pebbles from Permian conglomerate found at this level.

During Würm 3, a great reduction in bed-load transport energies is evidenced by the sandy-pelitic sequences described above. Peat-bog deposits are another sign of low-energy discharges, with spatially and chronologically distinct events of channel migration, oxbows and swamps formation, which must have persisted for some thousands of years. A persistence of cold climatic conditions is evidenced by fossil pollens found in the peat; on the other hand, a cold period is reported to have set in in Western Europe, roughly between 26'000 and 31'000 y BP (LEROI-GOURHAN, 1974; CHAMLEY, 1974).

Between the latest Würm and the Middle Holocene, a marked break in the sediment deposition seems to occur in the area corresponding to the present-day alluvial plain of the Po River. This perhaps could be seen in relation to stream activity characterized by erosional processes prevailing on aggradation. Stratigraphic gaps in the Holocene fluvial deposits are likewise reported to have occurred elsewhere, e.g. along some reaches of Brenta River, Northern Italy (CASTIGLIONI & PELLEGRINI 1981), as well as in some areas of Central Europe (STARKEL, 1982; LOZEK, 1984). The broad-sized deposits characterizing the lower part of Po Valley Holocene sediments suggest the re-establishment of a braided-like pattern connected to high-energy discharges, which probably persisted until historical times, as suggested by the artifact remains commonly found. The wealth of buried logs and artifacts incorporated in these sediments, also witnesses a very active stream evolution, with recurrent catastrophic bank erosion events.

Later on, in the last few centuries a new change in the Po fluvial pattern occurred, leading to the present-day irregularly meandered floodplain with prevailing overbank deposits.

4 CONCLUSION

Based on the hypothesis that the top of the sandy-pelitic complex, where present, may be considered as the lower limit of Holocene, the research at C.N.R.-I.R.P.I.-Institute is being developed about radiocarbon ages of organic remains contained both in the peaty horizon and Holocene deposits. The main attempt is to ascertain the extent of the supposed gap in sedimen-tation in several sites along the Po River alluvial plain South of Torino between late Würm and Early Holocene, as well as to establish an absolute time-scale for the Holocene so as to give quantitative information on sediment transport capacities during that period.

5 REFERENCES

ANSELO, V. & TROPEANO, D., 1978: Eventi alluvionali nel bacino del Torrente Banna (Torino) con speciale riferimento alla pienca del 19 febbraio 1972. Boll. Ass. Min. Subalpina, 15(3), 473-503.

BORTOLAMI, G., MAFFEO, B., MARADEI, V., RICCI, B. & SORZANA, F., 1976: Lineamenti di litologia e geoidrologia del settore piemontese della Pianura Padana. Quaderni dell'Istituto di Ricerca sulle Acque, 28(1), 1-37, Tip. G., Bardi, Roma.

CARRARO, F., 1976: Diversione pleistocenica nel deflusso del Bacino Piemontese meridionale: un' ipotesi di lavoro. Gruppo di studio del Quaternario padano, Quaderno no. 3, 89-100, Litogr. Massaza & Sinchetto, Torino.

CARRARO, F., PETRUCCI, F. & TAGLIAVINI, S., 1969: Note Illustrative della Carta Geologica d'Italia, F. 68 (Carmagnola), 1-40, Serv. Geol. It., Roma.

CASTIGLIONI, G.B. & PELLEGRINI, G.B., 1981: Geomorfologia dell'alveo del Brenta nella pianura tra Bassano e Padova. In: Provincia di Padova, Università di Padova: Il territorio della Brenta, 12-42.

CHAMLEY, H., 1974: Place des argiles marines parmi divers indicateurs paléoclimatiques. Colloques Internationaux du C.N.R.S., n. 219, 25-37.

GABERT, P., 1962: Les plaines occidentales du Pô et leurs piedmonts. Etude morphologique. 1-531, Imprimerie Louis Jean, Gap.

GASTALDI, B., 1872: Cenni sulla costituzione geologica del Piemonte. Boll. R. Comit. Geol. It., 3, 14-32; 77-96.

KAROL, R.H., 1960: Soils and soil engineering, 1-194, Prentice-Hall Civil Engineering Series, N.M. Newmark, Edit.

LEROI-GOURHAN, A., 1974: Analyses polliniques, préhistoire et variations climatiques quaternaires. Colloques Internationaux du C.N.R.S., n. 219, 61-66.

LOZEK, V., 1984: Late Glacial and Holocene development of Bohemian River Valleys. Abstract of paper presented at the IGCP Symposium 158 - A "Palaeohydrology of the temperate zone in the last 15'000 years", Mikulcice, Czechoslovakia.

MARAGE, F., 1983: Morphologie fluviale et migration des cours d'eau dans la haute plaine du Pô (Italie, partie Nord-Ouest). Geol. Jb., A 71.

PEOLA, P., 1942: Influenza dell'espansione glaciale sull'evoluzione del Fiume Tanaro. Boll. Soc. Geol. Itl, 61, 366-388.

PETRUCCI, F. & TAGLIAVINI, S., 1968: Considerazioni geomorfologiche sul settore occidentale del bacino fluvio-lacustre villafranchiano di Villafranca d'Asti (Quaternario continentale Padano-Nota 2). L'Ateneo Parmense, 4, 1-32.

PREVER, P.L., 1907: I terreni quaternari della Valle del Po dalle Alpi Marittime alla Sesia. Boll. Soc. Geol. It., 26, 523-556.

SACCO, F., 1917: L'evoluzione del Fiume Tanaro durante l'era Quaternaria. Atti Soc. It. Sc. Nat., 56, 157-178.

STARKEL, L., KLIMEK, K., MAMKOWA, K. & NIEDZIALKOWSKA, E., 1982: The Wisloka River Valley in the Carpathian foreland during the late Glacial and the Holocene. Polish Academy of Sciences, Geographical Studies, Special Issue No. 1, Evolution of the Vistula River Valley during the last 15'000 years, Part I, 41-56.

TROPEANO, D., ARDUINO, L., BOSSO, C. & FORNARO, M., 1984: Il Pliocene di La Loggia (Torino). Riv. Piem. St. Nat., 5, 55-67.

TROPEANO, D. & CHERCHIO, E., 1984: L'orizzonte torboso würmiano nel sottosuolo della Pianura Piemontese meridionale. Boll. Ass. Min. Subalpina, 21(3), (in press).

3 Fluvial environments and palaeohydrology: The state-of-the-art
Two views from Poland

Lake, Mire and River Environments, Lang & Schlüchter (eds)
© 1988 Balkema, Rotterdam. ISBN 90 6191 849 9

Valley floor development and paleohydrological changes: The Late Vistulian and Holocene history of the Warta River (Poland)

S.Kozarski, P.Gonera & B.Antczak
Quaternary Research Institute, Adam Mickiewicz University, Poznań, Poland

ABSTRACT

Results of a case study on the Warta River are presented in this paper which demonstrate climatically controlled valley floor evolution and paleohydrological changes.

Three low terraces, i.e. bifurcation, transitional and floodplain, reveal distinct paleochannel scars: braided, large paleomeanders and small paleomeanders respectively. Well dated sequences of paleochannel fills show differences in the age of channel pattern as well as vegetation changes in the drainage basin. The latter generated threshold conditions in the system responsible for varying run-off and discharge, thus the changing channel pattern and paleomeander geometry reflect paleohydrological conditions.

The paleohydrological reconstruction in the study area was based on : (1) detailed investigation of former bed load (transport zones) in the paleochannel, (2) measurements of bed load in the present-day channel related to velocity distribution and (3) estimation of paleovelocities with the application of empirical formulae.

By means of the above procedure a general paleohydrological tendency was found showing a decrease of discharges after the Bølling Interstadial until the end of the Subboreal Period from 263.9 m^3s^{-1} to 51.7 m^3s^{-1}. It is explained by growing density of vegetation in the drainage basin correlated with plant recolonization after the recession of the last ice sheet.

1 INTRODUCTION

After the recession of the last ice sheet, the rivers of the North Polish Plain were subject to the dramatic change in channel pattern: from braided to sinuous and/or meandering (KOZARSKI & ROTNICKI 1977). The change in channel pattern meant also a change in valley floor development. Very flat and broad sandy-gravelly bottoms formed by braided rivers were cut by better organised meandering rivers which produced a new valley floor morphology: multi-concave scarps, aban-

doned meander loops and point bar ridges. The river metamorphosis and channel changes were supposed to be a system response to climatic fluctuations. Therefore for detailed studies a test area in the valley of the Warta River to the south of Poznań was chosen in 1977 (Fig. 1) which was included in the IGCP 158 Project 'Paleohydrology of the temperate zone in the last 15 000 years' (KOZARSKI 1983a, b).

There were two opportunities to present progress reports and articles on the study discussed in this paper. First, in 1981, during the IGCP Project 158 (Subprojects A and B) International Symposium held in Poznań, Poland (KOZARSKI 1981a, 1983a). Second, in 1983, at the Severn'83 Symposium, Shrewsbury, England (KOZARSKI 1983b). Both reports did focus mostly on research methods of paleochannel scars and their fills, as well as on dating problems and the age of channel changes. Less attention has been paid to paleo-

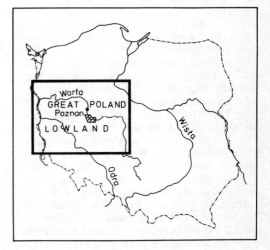

Figure 1. Location of the study area.

hydrological problems because only preliminary interpretations of general paleohydrological tendencies have been presented (KOZARSKI 1981, 1983a, b).

The present paper refers to: (1) geological setting of the valley floor related to the deeper substratum, (2) depositional models of terraces and new age determinations of paleochannel fills and (3) quantitative paleohydrological estimates in reference to climatic fluctuations during the Late Vistulian and Holocene. Results obtained in the case study during the last two years bring the Warta Valley Research Project closer to the guide-lines defined by STARKEL & THORNES (1981) for key-sites in the IGCP Project 158 A 'Fluvial environment'.

2 GEOMORPHOLOGICAL AND GEOLOGICAL SETTING

As was stated in earlier publications (KOZARSKI 1981, 1983a, b, KOZARSKI & ROTNICKI 1977) the bottom of the Warta valley in the case study area is formed by three levels: the bifurcation terrace, the transitional terrace, and the flood-plain. On every terrace paleochannel scars can be easily identified on aerial photographs, topographical maps and in the field. The scars have the form of braided channels on the bifurcation terrace (66-65 m a.s.l.), fragments of great meander channels on the transitional terrace (66-61 m a.s.l.), and fragments of smaller meander channels on the floodplain (60-58 m a.s.l.). Differences in the channel pattern especially of the hydraulic parameters are related to variable discharge and to the evolution of the valley floor over time. It was demonstrated (KOZARSKI 1981, 1983a, b) that the processes took place in the Late Vistulian and Holocene. The most recent data (ANTCZAK 1985, GONERA 1984, KOZARSKI, ANTCZAK & GONERA 1984), to be discussed in the subsequent part of the paper, support the claim presented above.

The bifurcation terrace west of the area under study occurs as a continuous level and constitutes the lowest level of the Warsaw-Berlin Pradolina, whose maximum width at this place does not exceed 8 km. So far, traces of SE-NW oriented braided channels have been found only at this fragment of the Pradolina. The main channel consists of three great branches of the mean length being 7790 m (ANTCZAK 1985). The channel depth has been estimated on the basis of the mean depth (i.e. 1.42 m) at which the shallowest channel lag deposits appear. The latter is the effect of the last flow on the terrace surface. The channels are separated by large sand accumulation surfaces in the form of bar groups from 5712 to 1050 meters long (2995 m in average), and from 837 to 392 meters wide (610 m in average). Five such large surfaces have been distinguished (ANTCZAK 1985). Following the last flow phase on the bifurcation terrace, the sand accumulative surfaces were cut by the system of smaller channels which,

together with the branches of the main channel, form a common system. The depth of the smaller channels is similar to that of the main channel, the length equals 1282 meters and the mean width equals 543 m. The bars between smaller channels are from 1200 m to 80 m long and from 0.4 to 1.3 m high. The latter are often greater when covered by eolian sands on the bar surface (ANTCZAK 1985).

Morphometric measurements both in the field and on aerial photographs permitted a determination of geometric parameters of paleochannels and a calculation of basic indicies of the channel pattern. The index of sinuousity and the index of braiding after BRICE (1964) have been calculated. The index of braiding expressed by the formula:

$$B_1 = \frac{2\,(Lb)}{Lc}$$

where (Lb) is the length of bars in the channel reach under study and (Lc) is the length of the channel measured along the morphological axis. This is 8.36 for the reach of the braided channel under study. According to MOODY-STUART (1966) the index shows that channel deposits exceed overbank deposits.

The index of sinuousity for the main channels has also been calculated. The mean value of the index is 1.065. Hence, these are channels characterized by low sinuousity, typical of braided channels.

The reach of the channel discussed here constitutes a greater braided system whose deposits were eroded in the course of later development phases of the Warta valley. According to MOODY-STUART (1966), the morphometric features are typical of braided channels which tend to aggrade. The high value of BRICE's (1964) index of braiding, i.e. = 8.36, indicates a great dynamics of the channels, which is a result of the hydrological regime's variability as a braided river.

The analysis of the topography and planer geometry of paleochannels at the transitional terrace and floodplain should be preceded by a few general remarks on the division of the paleomeander pattern into an older and younger set.

Geomorphological causes which justified the division of the observed paleomeanders, or, to be exact of their fragments into two groups (KOZARSKI 1974a, KOZARSKI & ROTNICKI 1977, 1978) were detected in early stages of the investigations.

(1) A study of aerial photographs permits detection of differences in the size of the paleomeanders. Larger paleomeanders occur far from the present Warta river whereas smaller paleomeanders occur closer to it.

(2) Paleomeanders occur at two levels. The broader ones are characteristic of the transitional terrace whereas the smaller ones are typical of the floodplain.

(3) The transitional terrace and the floodplain are separated by a low edge (2-4 m) which runs between broader and narrower paleomeanders.

For that reason two groups of paleomeanders were considered as evidence for two different morphodynamic generations (KOZARSKI 1974a, KOZARSKI 1976, KOZARSKI & ROTNICKI 1977, 1978) and models for a difference in age were put forward.

Originally, two geometric parameters were measured and calculated (KOZARSKI 1974a, KOZARSKI & ROTNICKI 1977, 1978) in order to describe paleomeanders. These were: the mean radius of curvature (R_c) and the mean channel width (W), i.e. two parameters of planar geometry. Later, as investigations of deposits filling the paleomeanders developed and the number of test borings increased, a third parameter was determined (ANTCZAK & KOZARSKI 1981, KOZARSKI 1981c, d). The mean channel depth was treated as an information on the thickness of the paleochannel fill. For that reason this parameter should be treated as a first approximation and its accuracy will increase along with further borings. The parameter was assigned proper significance in paleohydrological reconstruction by GONERA (1984) who determined it as the channel depth (D) during the backfill stage. It is obvious that the corrected value of the depth will necessitate a change of the relation of channel width to channel depth (F) and to channel cross-section (A) which will be followed by changes in the reconstructed discharges (Q). Measurements obtained so far (KOZARSKI 1983b, GONERA 1984) prove differences in the size of paleomeanders. The differences are also typical for paleomeander generations determined earlier (KOZARSKI 1974). They also imply the differences in discharges (Q) which shaped the river channel at subsequent developmental stages of the valley floor.

Paleomeanders which remained on the transitional terrace are characterized by geomorphological distinctness and they can be easily identified on aerial photographs, as well as in the field and on maps in the scale 1:10,000. They retained all features of meander channels, i.e. a steepness of the concave bank, a low-angled slope of the convex bank, as well as the legibility of accompanying forms like point bar ridges (BORZYSZ-KOWSKI 1971, KOZARSKI 1974b, KOZARSKI 1976, 1981a, 1983b, KOZARSKI & ROTNICKI 1977, ANTCZAK 1978, GONERA 1984) and in some places, like chutes. Point bar ridges are especially significant for the topography of the entire system and for the transitional terrace. They represent systems of arch-like forms from a few to 6 m above the level of paleomeander fill and, as is evidenced by a good example of such a set of forms at Jaszkowo (ANTC-ZAK & KOZARSKI 1981) they are related to the deepening of the channel during its lateral migration. The situation at Jaszkowo (KOZARSKI 1974b), has been used as one of the elements to reconstruct valley floor evolution during the lateral migration of the meandering channel (KOZARSKI & ROTNICKI 1977, 1978, 1983).

Measurements of geometric parameters of the older meander system in the entire fragment of the Warta valley between Nowe Miasto and Mosina were attempted twice (KOZARSKI 1974b, 1983a, b). They were done on aerial photographs first and, at the second stage (KOZARSKI 1983a, b) on the basis of additional field measurements at sites selected for detailed geological investigations. The measurements of the second stage (KOZARSKI 1983a, b) are naturally more precise. The results are that the radius of curvature (R_c) and the channel width (W) of the older meander system are greater than those of the younger system. The same is true of the mean channel depth and the channel cross-section (see Table 1).

However, the measurements, although satisfactory for the general description of differences in the geometry of the older and younger meander systems, are not useful in detailed studies aiming at a paleohydrological reconstruction. For such purposes field measurements are needed. They have been done at selected sites (GO-NERA 1984). They show differences compared to the value given in Table 1. This problem is discussed later.

Similarly to broader paleomeanders also smaller ones are easily identifiable on the floodplain by the use of aerial photographs, on maps in the scale 1:10,000 and directly in the field. They form a charactistic element of the floodplain's relief with well preserved steep concave banks which often undercut sharply the level of the transitional terrace and thus cut large paleomeanders with a low edge. The occurrence of clearly developed point bar ridges whose relative heights do not exceed 1.5 m, inside the paleomeanders is very common. Compared to the older meander system the great number of paleomeanders and their considerable 'crowding' are striking features of the floodplain. At some places, e.g., near Czmoniec (KOZARSKI 1983c, 1981d) the 'crowding' forms a set of undercutting channels which are confirmed also by investigating their infill.

Smaller sizes of paleomeanders on the floodplain, perceivable already during routine analysis of aerial photographs and cartographic maps in detailed scales, were confirmed first (KOZARSKI 1974a, KOZARSKI & ROTNICKI 1977, 1978) measurements of the two parameters of planar geometry, i.e., the radius of curvature (R_c) and the channel depth (W). Results of the second stage in the investigation of planar geometry were identical with a simple description of the system (KOZARSKI 1983a, b) as shown in Table 1. The measurements performed at the Dabrowa Site (GONERA 1984) compare to those at the Czmoń, Mechlin, Gogolewko and Zakrzewice sites and are related to the older meander system: that in each case the radius of curva-

Table 1. Basic geometric parameters of paleomeanders (in m)

Generation	R_c	W	D	F	A
Older	247.1	58.9	3.64	16.2	107.2
Younger	140.6	45.0	2.17	20.7	48.8

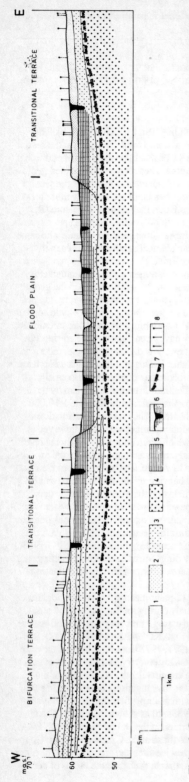

Figure 2. Geological cross-section of the Warta River valley (after ANTCZAK 1985). 1 - fine-grained sand, 2 - medium-grained sand, 3 - coarse-grained sand, 4 - gravel and pebbles, 5 - sandy-clayey cyclothems, 6 - organic paleochannel fills, 7 - erosional bed, 8 - bore holes.

ture (R_e) of the paleomeander on the floodplain is much smaller and occurs in the ratio of 1:2.2; 1:2.3; 1:3.0; 1:2.9. And the width is 1:0.8, 1:1.6; 1:2.4; 1:1.5; – it is smaller in three cases out of four.

A geological section of the Warta valley was constructed (ANTCZAK 1985) on the basis of archival materials from public work and on the basis of investigations during this study.

The sub-Quaternary substratum of the segment of the Warta valley under study is composed of Pliocene clays and Miocene brown coals. The substratum is at an average depth of -40 m to -50 m. In the western part of the profile the substratum rises to -10 and in the eastern part to -22 m.

Quaternary deposits occur as sands, gravels and tills. Tills occur here only insularly. Their presence is noticed under the surface of the transitional terrace and under the surface of the floodplain, however, there are no tills under the channel of the present-day Warta. The deposits reach the average thickness of 3-5 m.

To the west of the Warta channel tills reappear within the transitional terrace where they join the tills of the ground moraine surface.

Gravel and interlayered sands (Fig. 2) form the substratum of the actual Warta valley. A hypothetical line delimitating the extent of the pradolina substratum from valley deposits has been determined at the depth where coarse gravels with pebbles occur which, most probably, constitute the level of reworked tills. This level is in the Warta valley at a mean depth of 8-9 m. It is most shallow underneath the western part of the river channel (Fig. 2). The level is deepest where terrace surfaces meet; its maximum thickness there is 14 m (a contact between the surfaces of the transitional terrace and the floodplain). The mean thickness of fluvial deposits is from 8 to 10 m. These deposits rest on reworked tills. A similar procedure to separate deposits of the pradolina from valley deposits has been used by HANNEMANN (1961).

Fluvial deposits consist of gravels, sands and silts as well as organogenic deposits from paleochannel fills. According to KOZARSKI (1981a, 1983a, b), KOZARSKI & ROTNICKI (1977) three series of terrace deposits can be distinguished. The bifurcation terrace is built by a series from the end of pleni-Vistulian and the beginning of the Late Vistulian, the transitional terrace by Late-Vistulian sediments, and the floodplain by Holocene deposits. The sediments from the pleni-Vistulian and the beginning of the Late Vistulian are composed of sand and gravel and of small amounts of silts and organogenic deposits and forming, according to KOZARSKI & ROTNICKI (1977), the so-called basal series. Only at the Żabinko Site fine sands and silts with organogenic matters have been found in fills of braided channels. Deposits of the Vistulian are mainly fine sands and silts as well as organogenic deposits such as peats and gyttjas in the paleomeanders. The Holocene sediments are developed in a similar way, yet gyttja is not present there. The deposits of the Late-Vistulian and Holocene are also composed of eolian sands.

The analysis of the geological section of the Warta valley supports the following conclusions:
(1) In the study area a clear depression in the sub-Quaternary substratum can be seen (from -10 m in the western part of the valley to -15 m in the centre and -22 m in the eastern part of the valley), which might indicate that the valley was permanently present in this area. This problem has already been addressed by KOWALSKA (1957) and KRYGOWSKI (1961) who pointed to the existence of a depression in the cretaceous surface.
(2) The actual Warta velley developed in gravelly-sandy Pleistocene deposits filling the depression in the sub-Quaternary substratum. Reworked tills constitute the contact.
(3) The morphology of the actual Warta valley refers to the morphology of the sub-Quaternary substratum. However, the development of the terrace systems is not closely controlled by the relief of the substratum (Fig. 2).

3 DEPOSITIONAL MODELS AND AGE OF RIVER TERRACES

The actual morphological floor of the Warta valley is formed by low bifurcation and transitional terraces and the floodplain. These are aggradation terraces genetically connected to the lateral migration of the meander channels and to the deposition and shallow cutting of the transitional terrace and floodplain. A model proposed by KOZARSKI & ROTNICKI (1977, 1983) has been adequate so far as to explain the processes of deposition and undercutting. The aggradational character of the terraces and floodplain requires that attention be paid to the terrace deposits as they contain the paleohydrological record.

Depositional models are based on information from selected summary sequences as shown in Fig. 3. They comprise the better studied (ANTCZAK 1985, KOZARSKI, ANTCZAK & GONERA 1984) terraces of the western part of the study area and are in accord with the field survey on several sequences in the eastern part. As for the bifurcation terrace considerable information is from the important Żabinko-site with it's fill of the braided channel. In the case of the transitional terrace and the floodplain we refer to earlier publications (KOZARSKI 1981, 1983a, b, c). For that reason Fig. 3 does not include oxbow lake facies.

The typical profile of alluvium of the bifurcation terrace contains the following sequence (from bottom to top, Fig. 3):
– Large scale cross-bedding of the trough or tabular type,
– small scale cross-bedding,
– sediments of the paleochannel fill,
– covers of structureless eolian deposits.

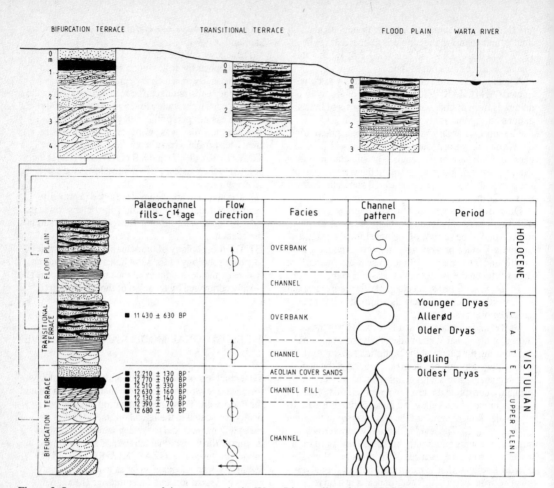

Figure 3. Summary sequences of river terraces in the Warta River valley after ANTCZAK 1985 with some changes).

This sequence reflects the variable hydrodynamic conditions from the high-energy environment to a complete disappearance of flow.

Sand and gravel of the first two beds occur as extensive lenses, several meters long. Silt forms the top of channel fills or of sand-silt rhythmites. At the Żabinko-site rhythmites build up the top of the terrace. This is evidence for flood stages, periodically iterative, in the decay phase of the flow at the terrace level. The depositional sequence in braided channels is as follows: silt, fine-grained sand, coarse- and medium-grained sand, gravel, channel lag-deposits (from top to bottom). This is a single cyclothem typical of a braided river (ALLEN 1970). Often the top layers are reduced due to lateral erosion as evidenced by channel lag deposits. Such an iterative sequence of deposits has been discovered by drilling the top of the bifurcation terrace to a depth of 6 m. The deepest braided paleochannels at Żabinko contain organogenic deposits, peat and gyttja.

Three hundred and twenty four samples of mineral deposits have been analysed for grain-size composition, the degree of abrasion and the calcium carbonate content. The variability of statistic indicies of grain-size composition after FOLK & WARD (1957) and FRIEDMAN (1962) and of the abrasion degree allow a distinction of the following terrace facies associations:
(1) Coarse-grained deposits of the channel facies
(2) Fine-grained deposits of the overbank facies, poorly sorted
(3) Fine-grained eolian facies, well sorted.

Sandy terrace deposits are carbonate free. Silt contains 5-6% of calcium carbonate on average. Considerable amounts of calcium carbonate, i.e., 30-40%, are found in gyttjas of the braided paleochannels. The analysis of flow directions in the aggradation series of a braided river shows that bifurcation of waters at this segment of the Warsaw-Berlin Pradolina ended at the level of the bifurcation terrace. A clear differentiation in the outflow directions has been found both in the vertical and horizontal pattern of sedimentation. In the vertical pattern there is a tendency to increase the northern direction towards the top (Fig. 3). The same tendency exists on the terrace surface for the north-south direction. But further conclusions are impossible because the bifurcation terrace is reduced in the southern part due to the erosion by the river's incision into the surface of the transitional terrace. A change in flow direction in the vertical profile is 50°-60° on average from the western to the northern sector. In lower terrace levels the northern direction is dominant.

The flow direction for the bifurcation terrace sediments shows a small scatter of data points. An average standard deviation of the flow directions for the channel deposits is 1.26 and for the point bar series it is 1.87. Such a small scatter of directions is typical for braided rivers (WILLIAMS & RUST 1969, SELLEY 1976).

An important section in sediments of the braided channel pattern which contains the key horizon is discovered in 1981 by B.NOWACZYK at Żabinko, where an extensive excavation can be found. The excavation permit to follow from 1981-1984, in changing intersections, the structures of an inland dune with a maximum height of 14 m. Also its base to the depth of 3 m was visible: fluvial sands, silts and peats (or organic matter similar to peat) are conclusive of a closed basin. Eolian sands will not be discussed here. But fluvial and basin deposits will be characterized in more detail.

The following profile (Fig. 4) represents an exposure of > 100 m in horizontal extension:

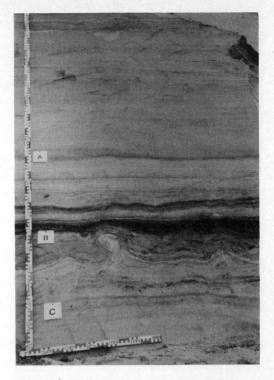

Figure 4. Żabinko. Key horizon in fill of braided paleochannel. The organic layer (B) was deposited during the Bølling Interstadial.

0.0-0.6 m	(A)	horizontally stratified, fine-grained river sands
1.6-1.7 m		horizontally stratified, fine-grained river sands, strongly reddish-brown due to a large concentration of ferric hydroxide
1.7-1.8 m	(B)	organic deposits; strongly decomposed and mineralized organic substances; overlying moss peat, weakly decomposed with numerous macro-remnants; the lower part being strongly decomposed and mineralized organic substance
1.8-2.0 m		greenish-grey silt (gyttja), greasy; lower part = greenish-grey silt rich in sand (gyttja); the silt layers display syn-sedimentary deformations down to the basal sand
2.0-2.25 m	(C)	fine-grained sand with silt, horizontally bedded; at two sites initial syngenetic frost cracks and marble structures related to the segregation ice of Late Vistulian permafrost
2.25-3.0 m		coarse- and medium-grained sand; with well developed small scale cross-bedding; flow directions to the NW; well developed synsedimentary

frost cracks, broadening to the top, with gravitational deformations on the sides.

Figure 4 gives a segment of a pit face with the sequence described but without the coarse- and medium-grained sand at a slightly greater depth at this site (N-S pit face in the eastern most part of exposure).

This profile shows that the sediments in the braided channel are recording events related to water flow and gradual accumulation and with the presence of permafrost. With declining flow, a shallow basin sedimentation is represented by fine-grained sand and silt (Fig 4c) and higher in the section by silt (gyttja) and organic deposits (Fig. 4b). The top of the sequence (Fig. 4a) with the horizontally bedded layer of fine-grained sand, is interpreted here as the result of sedimentation in a low-energy environment, i.e., in flowing water at very low velocities.

The organic deposits (Fig. 4b) have been ^{14}C dated. Samples were taken from a decomposed organic substance, from peat and pieces of wood retained at the base of the peat. The dates point to a Bølling Interstadial age, for the organic layer (Fig. 3). The oldest samples are the pieces of wood with Gd-924 = 12 770 ±

190 and Gd-1701 = 12 680 ± 90 y BP.

The typical sequence of deposits of the transitional terrace, with the exception of paleomeander fills, is as follows (from bottom to top):
– coarse- and medium grained sand with large scale cross-bedding, mainly of the trough type,
– fine-grained sand with small scale cross-bedding or covers of flood rhythmites or
– silt filling the swales,
– structureless fine-grained sand.

Deposits of the transitional terrace are characterized by a decrease in the coarse fraction. At the base of the terrace with the coarsest deposits (the so-called 'basic series' according to KOZARSKI & ROTNICKI 1977, 1978) large-scale cross-bedding of the trough type is observed (Fig. 3). The beds are up to 1 m thick and extend for 3-4 m on average. Within these sediments – similarly to a higher terrace – a fining upward sequence is observed. FRAZIER & OSANIK (1961), HARMS & FAHNESTOCK (1965), ALLEN (1968), MCGOWERN & GARNER (1970) associate the genesis of these structures with the migration of meander point bars.

These forms occur in general on the surface of this terrace and are connected with the accumulation due to lateral migration of the meander channel. The reduced thickness of the cross-bedding sets as compared to those of the bifurcation terrace, is due to a change in the river activity: from aggradation at the bifurcation level to erosion associated with lateral accreation of deposits typical of meandering rivers (KOZARSKI 1981a, 1983a, b). Finer deposits, i.e., fine sand with small scale cross-bedding or fine sand-silt rhythmites overlie the large scale cross-bedding. Climbing ripples or flaser bedding dominate the fine-sand sequences. The unit's average thickness is 70-100 cm.

Horizontal bedded layers usually intercalate with small and large scale cross-bedding. On average they are of a small thickness reaching 30-40 cm. The genesis of this bedding was discussed by UNRUG (1959), ALLEN (1964), PICKARD & HIGH (1973) and RAY (1976). The authors claim that this type of structure has not been explained yet. Horizontal bedding of deposits is, in those authors' opinion a result of the deposition from 'clouds' of non uniform density load varying in grain-size, transported by a constant, quiet, not too fast flow.

At the top of the terrace up to 1.5 m thick fine structureless sand occurs. The lack of sedimentary structures at the macroscopic-scale results from reworking by wind or from the obliteration of the primary structures by post-sedimentary processes, such as thawing and freezing in an annual cycle, or by infiltration of water or by bioturbation.

The alluvial deposits of the floodplain are difficult to study due to unfavourable hydrological conditions. They are connected to the deposition of a meandering river and to the lateral migration of channels of the younger generation of paleomeanders (KOZARSKI

1974a, 1976, 1981a, 1983a, b, KOZARSKI & ROTNICKI 1977, 1978).

These deposits consist mainly of fine-grained sand with small and large scale cross-bedding and sporadically, closer to the top of the terrace – with horizontal bedding. Small scale cross-bedding and flaser bedding, characteristic of a low-energy environment, are most frequent. Within point bar ridges and levees structures resulting from the migration of climbing ripples are dominant whereas structureless silts or showing traces of flasers, are fairly frequent in swales.

The rhythmites in the top of the floodplain sequences are widely distributed and will be discussed in the next chapter.

The typical sequence in the floodplain deposits is as follows (bottom to top):
– medium-grained sand with large scale cross-bedding,
– fine-grained sand with small cross-bedding or
– rhythmite covers and
– silt filling the swales.

Rhythmite covers of repetitive cycles of sandy-silts are widespread on the surface of the transitional terrace and the floodplain. The top layers of rhythmically stratified deposits were investigated in the transitional terrace and in the floodplain at 19 sites. On the transitional terrace these top units are restricted to an island. But on the floodplain they make an almost continuous cover.

The rhythmites are of a variable thickness from 0.4 to 3.2 m in the transitional terrace and up to 2 m in the floodplain. The rhythmites are developed as extensive sandy-silty lenses of a small thickness. The lenses are from 2.3 to 12 m long and their thickness is from 3 to 24 cm. Within the thickest silt layer a characteristic sand layer occurs. Such layers are of a few millimeters to 2-3 cm only.

Individual sets of rhythmites are characterized by repetitive occurance of fine sands and silts. Contacts between the sets are clear and sharp as opposed to a gradual transition between sandy and silty deposits within the set. This feature indicates a gradual decrease of flow during the flood period which caused the deposition of a fining upward sequence. The following flood began with erosion or with deposition of the coarsest deposits.

The sand layers of the rhythmites are generally structureless. In the thickest sand layers traces of small-scale cross-bedding is noticed. The deposits are related to periods of flow on the terrace surface. Some contacts display deformations at the contacts of sandy and silty layers with a character of microfolds. Such deformations are related to density inversion and instability of genetically different deposits. The deformations occur at depths greater than 2 m and are absent in cyclothems of shallower origin.

Variable colour from light yellow in sand to brown in silt is a characteristic feature of these deposits.

In order to describe quantitatively the variability of rhythmites and to compare the character of floods which created them (ANTCZAK 1985) a simple index (P/M) was calculated. The index is a quotient of the thickness of a sandy layer (P) and silty layer (M). The value of the index provides information on the scale of deposition during a flood event on the terrace and during the later stagnation phase i.e., on the scale of deposition from suspension and decantation.

In the sequence of the transitional terrace under study the index is in the range of 0.49-3.82. Its value increases towards the bottom as a result of the accumulation from suspension. The entire thickness of cyclothems displays the same tendency. These regularities imply that the older cyclothems were produced by higher flood stages.

The decrease of thickness in individual sets of rhythmites is also observed further away from the paleochannels. At the transitional terrace they disappear completely. In any case this is the effect of paleohydrological variability of the river as well as of local accumulation conditions caused by the migration of the meander channel and the morphology of the terrace.

The top layers of rhythmites on the floodplain show analogous tendencies to those of the cover in the transitional terrace. The value of the P/M index increases towards the bottom of the terrace. The value is in the range of 0.26-2.33 and the thickness of individual sets is from 3 to 18 cm. This regularity shows that both the flood amplitude and the resulting sediment accumulation were greater in the transitional terrace.

Geomorphological, geological and paleobotanical facts as well as a progress in the dating of paleochannel fills (Fig. 5) allow for a more precise chronology of events in the Warta river valley and for a determination of the age for the bifurcation terrace, the transitional terrace and the floodplain. They also initiate the discussion on a quantitative treatment of paleohydrological problems in the next chapter. These are:
(1) The problem of river evolution and changes in its channel pattern as a response to climatic changes
(2) The general paleohydrological trend over time from the Upper pleni-Vistulian to Holocene, inclusively.

Geomorphological and sedimentary observations as well as the dating of the paleochannel fills prove that the river evolution and changes in the channel pattern were events which occured over a considerable time span. Results obtained so far allow for a reconstruction of the following sequence of events:

Upper Pleni-Vistulian and the Oldest Dryas	braided channel (bifurcation terrace)
Late Vistulian from the Bølling Interstadial to the Younger Dryas, inclusively	meandering channel – large paleomeanders (transitional terrace)
Holocene	meandering channel – small paleomeanders (floodplain)

Boundaries between the periods of changes in the river channel pattern are clearly determined by paleobotanic analyses and/or measurements of the age by means of the radiocarbon method. In a chronostratigraphic schema for the Great Polish Lowland (KOZARSKI 1981b, 1983b). The first boundary occurs between the Oldest Dryas and the Bølling Interstadial. It is manifested by an organic bed at the top of the sandy and later silty fill of the braided channel which was deposited during the Bølling Interstadial (Figs 3 and 5). This organic layer contains well preserved pieces of wood which demonstrates that this formation is in situ. Obviously, the braided channel had to be abandoned by the river when clastic deposits such as silt and organic deposits, i.e., deposits of a closed basin were accumulating. It implies that the braided channel was active during the Oldest Dryas and in the upper pleni-Vistulian after the Poznań phase of the Last Glaciation. The position of the second boundary is inferred from the age of the basal filling of large and small paleomeanders (Fig. 5) are of the Younger Dryas and the oldest associated with the Boreal Period (OKUNIEWSKA & TOBOLSKI 1982). Hence the boundary falls at the turn of the Late Vistulian.

Both boundaries of changes in the channel pattern coincide with the main periods of changes in vegetation cover. This is not accidental but is directly related to the hierarchy of changing vegetation (SCHUMM 1977). Type and density of the vegetation are dominant factors and they reflect climatic and edaphic conditions in a complex way. In paleogeographic reconstructions therefore these factors seem to make the satisfactory basis for explaining the river's response and changes in its channel pattern. A discussion of the paleovegetation cover can replace estimates on paleotemperature, precipitation, transpiration or speculations on the ratio between outflow and paleoprecipitation (KOZARSKI 1983b). Recently FRENZEL (1980, 1983) pointed out the difficulties involved. FRENZEL (1983) did demonstrate that it is the changing vegetation cover and not the water balance that caused the fluvial pattern to vary.

During the Upper Pleni-Vistulian, mostly without vegetation cover, and during the Oldest Dryas when a shrub tundra developed – as evidenced by pollen diagrams (TOBOLSKI 1966, BORÓWKO-DŁUŻAKOWA 1969) – the braided river pattern was typical. Convincing evidence of this was found at the Żabinko Site. Outflow, discharge and supply of sediment to the channel were at that time strongly dependent on parmafrost in the substratum. Hence, sheet flow and through flow were dominant and there was no possibility of infiltration (KOZARSKI 1983b). An intensive sheet flow removed material from vegetation-free (or almost vegetation-free slopes) to river channels. However, the main amount of the sediment load of rivers came from the lateral undercutting of noncon-

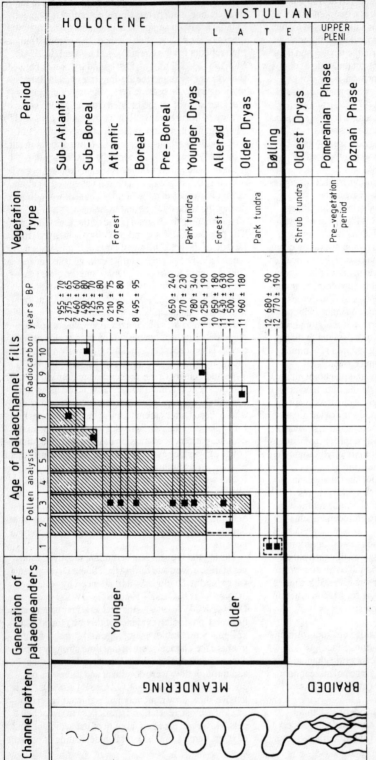

Figure 5. Chronology of the Warta River channel changes (after KOZARSKI 1981, 1983a, b, c and KOZARSKI, ANTCZAK & GONERA 1984). Local names of cores taken for paleobotanical (hachured logs) and radiocarbon dating (unhachured logs and squares): 1 - Żabinko, 2 - Mechlin, 3 - Jaszkowo, 4 - Zbrudzewo I, 5 - Zbrudzewo II, 6 - Czmoniec A, 7 - Czmoniec B, 8 - Gogolewko, 9 - Czmoń, 10 - Dąbrowa.

solidated glacial and fluvioglacial deposits. In the presence of permafrost the latter deposits, being noncohesive, are especially liable to undercutting, which is evidenced by actualistic studies in Alaska (SCOTT 1978). In the case of a braided river with tendencies to lateral erosion the sediment yield from undercutting had to be especially important. It is also reported from actual periglacial area (CHURCH 1972) that the sediment transport is 90% of bed load. It was pointed out earlier that this material as coarse-grained sand and fine gravel is characteristic of abandoned channels in our study area. According to SCHUMM (1968) channels carrying a coarse bed load are characterized by a large ratio of width to depth. In the Warsaw-Berlin Pradolina this statement is evidenced by both shallow and wide paleochannels as well as by a considerable (8 km) extension of a bifurcation terrace which was formed during the Upper pleni-Vistulian and during the Oldest Dryas by a braided river.

A fairly rapid change in the vegetation type after the Oldest Dryas initiated allogenic changes in the river channel (LEWIN 1977). A transition from a poor shrub tundra to park tundra in the Bølling Interstadial indicated that as threshold condition appeared (BRUNSDEN & THORNES 1979). Under these conditions the so-called extrinsic threshold (SCHUMM 1977, 1979) was exceeded, which caused the river to change from braiding to meandering. This was the most significant change in the channel patterns during the entire time interval discussed here in which climatic changes and responses in the river channel took place. After the changes during the Bølling Interstadial the river remained meandering up to the present time and man-made channel control.

It follows from earlier analyses (KOZARSKI 1983a) that in the middle segment of the Warta valley a general hydrological trend of decreasing average discharge (Q) was taking place over the entire period from the Upper Pleni-Vistulian to the Subboreal. A similar trend occurred in many lowland rivers of Poland (KOZARSKI & ROTNICKI 1977, SZUMAŃSKI 1983). These observations are obviously of a qualitative character. However, it is demonstrated in the next chapter that this survey is the background for an advanced paleohydrological modelling.

4 PALEOHYDROLOGICAL ESTIMATES

The general paleohydrological trend as discussed above is only a qualitative picture of the control mechanisms in river channel regimes. Changing concepts in paleohydrology, from qualitative to quantitative models, have been inspired by the fundamental reconstruction of 'hydraulic geometry' (LEOPOLD & MADDOCK 1953). The reconstruction was done on the basis of earlier investigations on paleochannel geometry and on the analysis of grain-size composition of channel bed deposits at the sites of Czmoń, Mechlin,

Gogolewko and Zakrzewice in the transitional terrace and at Dabrowa in the floodplain and near Bystrzek in the actual Warta meander.

The flow velocity in these paleochannels was calculated on the basis of the 'Manning formula', used in applied hydrology and of increasing use in paleohydrology (PARK 1978, KOUTANIEMI & RONKAINEN 1983, MAIZELS 1983, ROTNICKI 1983):

$$V = \frac{H_R^{2/3} \cdot S^{1/2}}{n}$$

This formula was re-arranged for permanent flow and ideal channel geometry. This is a considerable restriction for calculations on natural channels. But a good convergence of the measured and calculated velocities had added the 'Manning formula' to routine practice in hydrological services of many countries. The roughness coefficient contained in the formula was calculated after the 'Strickler formula' (CHOW 1959, GARDINER & DACKOMBE 1983):

$$n = \phi \left(\frac{H_R}{d} \right) d^{1/3}$$

$$\phi \left(\frac{H_R}{d} \right) = \frac{(H_R/d)^{1/3}}{21.9 \log (12.2 \, H_R/d)}$$

The application of this formula for flow velocity calculations in paleomeanders required that certain necessary assumptions be made. Since it was difficult to determine the slop of a channel for a given day, a complimentary method was used. It consisted in the measurement of the slope of the terrace on which the objects to be investigated can be found. The values obtained were close to the value of the slope of the actual Warta channel between Nowa Wieś Podgórno and Śrem (= 0.00018) and this value was accepted for further calculations.

Also the hydraulic radius was replaced in the above formulae by the depth defined for a 'full discharge' in a given paleomeander. This operation is possible (CHOW 1959) in a situation if the channel width is > 1/10 its depth. The depth of the bankfull discharge was obtained as follows: depth values given for individual borings were increased by a constant, calculated from the analysis of the paleochannel bank heights, while it was assumed at the same time that the bankfull discharge belongs to forming discharges (LEOPOLD, WOLMAN & MILLER 1964, ACKERS & CHARLTON 1970, PICKUP & RIEGER 1979, HICKIN 1981). The constant was 0.5 m at the Dabrowa Site, 1.5 m at Gogolewko and 1.0 m in other paleomeanders.

The slope angle of the banks in the investigated paleochannels changes on average from 30° for the erosional to 10° for the accumulative bank. If the constant factor was introduced to the calculations also the width values changed: with a constant of 0.5 m, the width increased by 3.4 m, with 1.0 m by 6.7 m and with 1.5 m by 10.0 m.

The application of the empirical Manning-Strickler

formula to calculations was preceded by field investigations on the actual Warta meander near Bystrzek. The usefullness of the Manning formula for the reconstruction of flow velocity in the actual Warta river should be tested. The convergence of mean measured with calculated velocities was determined.

The site is located on the 299th km of the river course, about 4.5 km east of Śrem. In this segment, the Warta river continues its parallel course to the Warsaw-Berlin Pradolina.

The basic geometric parameters of the meander are:
– the radius of curvature $R_c = 310$ m,
– meander wavelength $L = 1600$ m,
– amplitude $A_m = 800$ m,
– meander height $h = 400$ m,
– meander width $\lambda = 700$ m.

Field investigations were carried out at the river water level of 297 cm (water gauging station in Śrem), which corresponds to an average, throughout a period of many years, value of 290 cm. The measurements were made in 11 profiles at 100 m distance (GONERA 1984).

The average channel width was 60.7 m, varying from 52.0 to 69.0 m. The channel morphology is asymmetric. The greatest depths are near the concave bank. There, depth exceeds 3.0 m and at the meander outlet values > 4.0 m are reached. Towards the convex bank the channel becomes shallower. Only in profiles V and VI the depths approximate the value of 0.5 m (0.65 m and 0.53 m respectively). In the majority of cases the depth is about 2.0 m.

Measurements of flow velocity were done in 4 depths (V_{00}, $V_{0.8D}$, $V_{0.2D}$ and $V_{1.0m}$) at 61 hydrometric points. The velocities vary from 0.1 ms^{-1} to 0.9 ms^{-1}. Minimal values are noted at the channel bottom where only in a few cases the velocities exceed 0.4 ms^{-1}. In other depths the values increase to 0.8 - 0.9 ms^{-1}, rarely decreasing below 0.4 ms^{-1}. Comparing the distribution of mean velocities with depth we find that the highest and most frequently registered velocities in the order of 0.6-0.7 ms^{-1} exist at channel depths of 2.0 to 3.0 m. As the channel becomes either deeper or shallower the velocities decrease to 0.1-0.2 ms^{-1}. Moreover, the analysis of velocity profiles indicates that water flow is not uniform: currents with different flow velocities can be distinguished.

Sixty one samples of channel bed sediments for grain-size analysis were collected from the same places where flow velocities were determined. An Instorf-type borer specially adapted for sampling was used. A statistic description of actually transported load was made in laboratory analysis. The basal material consists of a medium-grained sand (96%), dominant in the depth range of 2.0-3.0 m and corresponding to the zone of maximum velocities. At the same time we noticed a tendency to a coarsening of the material along with the increase of the depth (to 0.35% at the depth of 4.45 m). With regard to sorting: a well sorted sediment dominates in the channel and it constitutes 41% of the

analysed material. In the spatial distribution of the standard deviation we noted a displacement of well and very well sorted materials towards the concave bank at the meander inlet and to the convex bank at the meander outlet. A reversed tendency is noticed in the case of moderately well and moderately sorted deposits. Poorly sorted material corresponds to the places of maximum depths.

These data enable a comparative analysis for mean measured velocities (mean of the sum 0.2D and 0.8D) and velocities calculated according to the Manning-Strickler formula. This analysis did not include data from measurements in the direct vicinity of the erosional bank, where because of the irregularity in the bank line of numerous whirlpools and backflows, very low values for the mean velocity were obtained. This decision is important because the Manning formula defines velocities for ideal flow conditions which did not occur at our sites. But our study is yet another warning for accepting calculated values in paleohydrological reconstructions as real figures. For the meander at Bystrzek the flow velocities were presented graphically. The statistical relationship between the values obtained was expressed by means of a power function with the power exponent of 0.5846. The correlation coefficient is $r = 0.4896$ and the application of the t student test (GREŃ 1984) confirmed the significance of the calculated relation at the significant level $t = 0.01$ despite a low determination coefficient (r^2) of 0.24.

Following the statistical calculations the Manning-Strickler formula is adapted to approximated natural conditions with

$$V_m = 0.7833 \, V_C^{0.5846}$$

and expanded with

$$V_m = 0.7835 \left(\frac{D^{0.67} \cdot S^{0.5}}{n} \right)^{0.5846}$$

The expanded form of the formula was used for the final estimates of velocity values in our paleochannels.

Another problem was the river's competence to transport the material of a given fraction as found by the grain-size composition analysis of channel bed deposits. The question is if a maximum, critical diameter of grains in transport can be related to a specific geometry of the paleochannel. Such calculations were done by using Shields criterion (GRAF 1971, DĄBKOWSKI, SKINIŃSKI & ŻBIKOWSKI 1982) which describes conditions of flow as expressed by a relationship between critical values of shear stress at the bottom (τ_c) and the grain-size (d). Studies by Shields have yielded a curve which defines conditions of tractable force at which bed particles begin to move, rather than the conditions at which all particles are in motion (MAIZELS 1983). Shields derived two dimensionless parameters representing two opposing forces: first, the resistance of the particles on the fluid, which is the particle Reynolds number and second, the drag force of the fluid on the grains expressed by the formula:

$$\phi = \frac{\tau c}{(\gamma_s - \gamma) d}$$

Using the above formula and assuming
(1) with Shields that critical shear stress (τ_o) equals the classical form of Du Boys' shear stress $(\tau_o$, SCHEIDEGGER 1984), $\tau_o = \gamma DS$ (kgm^{-2}); that
(2) $\phi = 0.056$ since this is the value most frequently accepted in the conditions of a turbulent flow (GRAF 1971, MAIZELS 1983), and that
(3) in channels with about 10% concentration of suspended material the specific weight of the suspension (γ) equals approximately 12 500 kgm^{-3}, then the formula for the critical diameter of the grain assumes the form:

$$d_c = \frac{\tau c}{769.1} \cdot 1000 \ (mm)$$

The critical grain-size defines the maximum size of the grain that begins to move when critical flow conditions develop where samples were taken.

The results from our analyses are (GONERA 1984) that the competence of the paleochannels under study is much greater than the size of material available for transport. The results obtained show that during conditions of the bankfull discharge we deal with a permanent transport of material the mode of which changes depending on varying hydrodynamic conditions in the channel.

Transport zones of channel bed deposits were distinguished on the basis of the statistical indices of a grain-size distribution such as mean size (Mz) and standard deviation (δ). The relationship between these two indices is important and instructive (SLY, THOMAS & PELLETIER 1983, KANIECKI 1976, KLIMEK & STARKEL 1981). It indicates that the dissipation of the particle size is not an entirely independent parameter but dependent on the average grain-size. Hence, we can speak of a dissipation greater than the majority of deposits with a given average diameter (GRZEGORCZYK 1970). In the case of the paleomeanders under study sandy deposits ranging from 1.55 to 2.75 ϕ have the best sorting. When the above values were marked on PASSEGA's C-M diagram (1964), it appeared that the individual types of transport are separated in a satisfactory way (SLY, THOMAS & PELLETIER 1983). So these values were accepted as boundaries for the discrimination of different zones in the paleochannels.

The investigation procedure with direct field studies and with using an empirical formulae did allow for a reconstruction of geometric parameters of paleochannels. If presented on a time scale, they show characteristic tendencies of change. The observed quantitative differences are a direct effect of the morphological activity of flowing waters which are governed by the laws of hydraulics. It should, however, be remembered that this activity depends on a number of variables such

as climate, geology and, in the recent past, human activity.

The basic assumptions for our analysis were the stratigraphy of the Late Vistulian and Holocene after ŚRODOŃ (1972) and MANGERUD et al. (1974), especially the bio- and chronostratigraphic aspects. Time analysis of the parameters mentioned above was possible through ^{14}C analyses, and, in the case of the paleomeander in Mechlin, also through paleobotanic analysis (OKUNIEWSKA & TOBOLSKI 1981). The results are supplemented in Fig. 6 by additional information on the limits of error in radiometric measurements. For the paleomeander in Zakrzewice where no ^{14}C dates on its stratigraphic position are available, a morphological criterion was used as this paleomeander is undercut by a dated paleomeander in Gogolewko. It should be explained that the dates obtained from the base of organogenic deposits determine the beginning of an oxbow lake in this abandoned channel. The lack of detailed studies to determine the beginning and the rate of eutrophication makes it difficult to date accurately the event of paleomeander cut-off. We have placed this event in the period directly preceding the obtained dates. Four of the investigated paleomeanders (Zakrzewice, Gogolewko, Mechlin, Czmoń) were dated for the period of the Late Vistulian and one (Dabrowa) for the end of the Subboreal Period. No doubt, the lack of sites from the beginning of Holocene made a more precise reconstruction more difficult.

The final analysis has aimed at a paleohydrological reconstruction and the following mean parameters of planar and hydraulic geometry were selected (Fig. 6): the radius of curvature (R_c), width (W), depth (D), the channel cross-section (A), velocity (V), discharge (Q) and river competence (stream power Ω). The last parameter, worked out by BAGNOLD (1966) is

$$\Omega = g \cdot D \cdot S \cdot V \cdot W \ (N)$$

It is a function for the size of both, the channel and discharge and as such is usefull in the statistics of river processes. Ω is directly proportional to the size of the river load (BAGNOLD 1966, GRAF 1983), and therefore somewhat difficult to be applied in paleohydrological reconstructions. For the investigated paleomeanders a selection of statistical indices for grain-size distributions (mean size and standard deviation) and a relative variance of the percentage of individual groups of fractions have also been presented. A dominant fraction in a given paleochannel has been specifically indicated.

A tendency to decrease the values for individual geometric parameters in progressively younger paleomeanders is generally noticed. Climatic changes in the Late Vistulian and Holocene, already discussed, should be considered to be the main causes of this phenomenon.

The oldest dated paleomeanders in the Warta valley south of Poznań, belonging to the older generation, is of the Bølling Interstadial. This time period is of a

197

special significance for the evolution of the river system in the middle reach of the Warta valley. This is connected with changes of climatic and related edaphic conditions in which a transition from shrub tundra to park tundra caused a threshold situation (HOWARD 1980) and a transformation of the channel pattern from braided to meander type (KOZARSKI 1983b). A paleomeander in the vicinity of Zakrzewice which belongs to the largest paleomeanders discussed in this paper. The presence of thick silt and very fine-grained sand, exceeding 2 meters, is the main feature which differentiates the paleomeander in Zakrzewice from the others. The origin of the deposit and its role in the formation of the channel cross-section remains to be explained. An answer to the question of origin is provided by investigations pertaining to the Oldest Dryas and Bølling, as given above. Severe climatic conditions, poor vegetation cover and the presence of permafrost favoured the sheet flow which supplied the channel with considerable amounts of sediment. Fine-grained material from the denudation of ground morain areas, in the direct vicinity of the river valley was dominant. The dominance of fine-grained sand in the paleochannel in Zakrzewice, whose mean is 47.5% of the entire population of grains, is such a consequence. An increase in suspension load has changed the properties of the environment, and has limited its erosional capability. Hence the size of the channel is smaller than could be inferred from the water balance which was positive at that time (FRENZEL 1983). A similar situation was found also in the bottom of the Gipping river valley, near Sproughton (ROSE, TURNER, COOPE & BRYAN 1980) at 12 200-11 300 y BP. The increase in the river load at that time is also mentioned by KLIMEK & STARKEL (1981). The discharge and velocity values, reconstructed for this site are 166.3 m³s⁻¹ and 0.74 ms⁻¹, respectively.

The Warta channel near Gogolewko functioned during the final phase of the Bølling Interstadial because we relied on the date of 11 960 ± 180 y BP (= Gd-2127) from the base of the paleomeander infill. This paleomeander has the highest values of selected geometric parameters. But with regard to the grain-size distributions it is in contrast to the site at Zakrzewice. The bulk of the material in the paleomeander at Gogolewko is medium-grained sand (51.4%) while the amount of coarse-grained sand is relatively large (7.1%). The presence of such material in the paleochannel at Gogolewko reflects the limited supply of fines from sheet flow due to an increasing density of the vegetation cover. At the same time, a 'cleaning' of the channel of fine-grained material was taking place. This resulted in the decrease of the cohesion that facilitated erosional processes, which had their effect in the maximum mean depth (3.8 m) and mean width (121 m). The paleomeander in Gogolewko also has the highest values for the 'stream power' (= 585 N), which, however, has not been utilized due to a limited maximum grain size. The reconstructed bankfull discharge (= 263.9 m³s⁻¹, Fig. 6)

exceeds the values found for the other sites several times. This is not true for mean flow velocity (0.72 ms⁻¹) which does not reach high values due to the roughness coefficient.

After the Older Dryas a second and longest and warmest climatic oscillation during the Late Vistulian, took place: the Allerød Interstadial. The last two sites covering the Late Vistulian time period, i.e., Mechlin and Czmoń, are correlated with the Allerød Interstadial. In these paleomeanders we observe a tendency to decrease the values of the analysed geometric parameters. At the Mechlin Site this is caused by a change in the relation of precipitation/run-off in favour of more balanced discharge with 127.8 m³s⁻¹. The denudation is limited by a dence vegetation cover and ameliorated channel process.

The deterioration of climatic conditions at the end of the Allerød preceded the Younger Dryas which was cooler and dryer as compared to the previous period (FRENZEL 1983). The smallest paleomeander at Czmoń is from that period. It was dated at 10 850 ± 180 y BP (= Gd-2129). For many features it is different from the paleomeanders discussed so far. The calculated values of flow velocity (Fig. 6) and bankfull discharge are 0.63 m³s⁻¹ and 51.8 m³s⁻¹, respectively, and are the lowest for the Late Vistulian. Low values of the mean depth (2.18 m) and width (45 m) and in most cases symmetric paleochannel cross-section indicate a reduced activity of channel processes. Again, a rapid increase in fine-grained deposits indicates the activation of sheet flow but which was limited by underground run-off. This was due to less dence vegetation, more of the park-tundra type (KOZARSKI 1981b). Dryer climate was also noticed and produced large-scale steppe vegetation (TOBOLSKI 1966). During the Youngest Dryas an increase in eolian activity was observed again. This was important in the Great Polish Lowland (KOZARSKI 1981b).

Another amelioration of climatic conditions took place at the beginning of the Holocene. The boundary between Late Vistulian and Holocene is marked in pollen diagrams by a rapid decrease in herbaceous pollen grains (NAP) and a simultaneous increase of birch and pine pollen (ŚRODOŃ 1972). According to KOZARSKI (1981b) a rapid expansion of forests at the beginning of Holocene caused the appearance of another threshold situation, important for the evolution of the river system in the middle reach of the Warta valley. The appearance of paleomeanders of the so-called younger generation on the floodplain is connected with that period. These forms display smaller paleochannels as discussed before. In order to find reasons for such a situation, we should trace the history of Early Holocene, especially with regard to a changing water balance. This should be done despite the scarcity of sites from that period.

Preboreal and Boreal were characterized by less precipitation, by less run-off and by an absence of extreme flooding (KLIMEK & STARKEL 1981). In Great Bri-

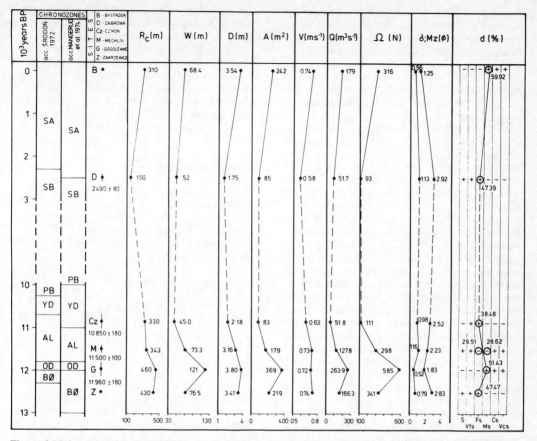

Figure 6. Main tendencies of changes in planar and hydraulic geometry and particle size of bed load at bankfull discharges during the Late Vistulian and Holocene in selected paleomeanders and the actual Warta River loop.

tain these were warm and rather dry periods favouring the development of thermophilous vegetation especially at the close of the Boreal (GOUDIE 1983). A further amelioration of climatic conditions during the Atlantic period as evidenced by an increase of temperature and humidity favoured the rise in river activity. The reactivation did not assume a similar size as during the Allerød. This was mainly due to a change in the species composition of the dense forest cover. The appearance of thermophilous deciduous trees, widely spread in the Atlantic period caused a rapid increase in transpiration, almost double actual values (FRENZEL 1983) and therefore decreasing overland run-off. One should not forget the role man played in the changing of the natural environment of that period (LIMBREY 1983). Following ŚRODOŃ (1972) the influence was limited during Paleolithic and the Mesolithic periods. Later, at the turn of the Atlantic and Subboreal periods the influence of Neolithic settlements became visible on pollen diagrams in the form of a clear and simultaneous decrease in the amount of trees which were part of

mixed deciduous trees.

The change was taking place during a steady deterioration of climate after the Holocene climatic optimum. The youngest and the smallest paleomeander in Dabrowa (Fig. 5), is dated at $2\,490 \pm 80$ y BP (= Gd-2125), and is conected to the end of the Subboreal period. With regard to geometric parameters it is similar to the Czmoń site, although their development courses differed. If the paleomeander in Czmoń is primarily of natural original the paleomeander in Dabrowa is influenced by human activity. Intensive agriculture caused a degradation of the forest areas, which increased overland run-off processes. This resulted in a considerable amount of fines (47.4%) in basal layers of the paleochannel. The bankfull discharge as reconstructed for this site is 51.7 m³s⁻¹ and almost five times smaller than the maximum as reconstructed for the Late Vistulian (Fig. 6). The mean flow velocity is 0.58 ms⁻¹.

Deteriorating temperature conditions from the end of the Subboreal Period throughout the entire Subatlantic until recent times and a simultaneous increase in humidity (ŚRODOŃ 1972, LAMB 1977) and rapidly developing settlements with the human impact on the environment caused more changes in the run-off pattern. Important was the decrease in transpiration (FRENZEL 1983). In the Warta valley such changes influenced the increase of the channel parameters. The increase was preceded by an increase in discharge to 179 m³s⁻¹ as compared to the end of the Subboreal Period (Fig. 6).

The understanding of the general trend in paleohydrological estimates as given here is only a small part of paleohydrological reconstruction which should aim at the most complete modelling of the river regime in the future. For example, the seasonal and multi year amplitude of fluctuations in the discharge volume and in river stages is one of the most important items of information to be gathered. Knowledge on the frequency of high flood stages as already mentioned is very important as well (KOZARSKI 1981a, 1983a).

For the upland areas of semi arid and humid temperate zones in the USA an interesting attempt at reconstructing flood stages on the basis of deposits accumulated in a low-energy environment (slack-water deposits) was presented by BAKER (1983). In the conditions of a lowland river the role of a similar information carrier may be played by flood deposits rhythmically stratified. They may permit an approximate determination of horizontal and vertical extent of the flooding. A study of cyclothems may provide data on the flood frequency. Preliminary interpretations along these lines from the study area in the Warta valley have been published by ANTCZAK (1985). A rhythmite unit of 3 meters in thickness on the transitional terrace is a potential site for such investigations since the occurrence of park tundra for known time intervals provides information on clearly unordered river regime, where large oscillations of water stages are typical.

5 CONCLUSIONS

Investigations within the Warta Valley Research Project contribute new results for:
(1) the determination of depositional models for low terraces and for the floodplain; discoveries of rhythmically stratified flood deposits were also important here;
(2) radiocarbon dating of paleochannel fills of the braided and meander types helped to close gaps in the chronology of events and to verify the age of the terraces and the floodplain; also paleohydrological modelling was put in a more precise time framework.
(3) quantitative estimates of paleodischarges and their changes during the Late Vistulian and, to a lesser degree, during the Holocene.

In attempting a reconstruction of paleodischarges we should first investigate the bed load distribution at the base of the oxbow lake facies.

Quantitative estimates of the paleodischarges in our study area show a clear decreasing tendency lasting from the Bølling Interstadial until the Subboreal Period. The problem of less discharge in the Bølling Interstadial requires further investigations since the paleomeander at Zakrzewice, as a reference for that estimate (GONERA 1984), was only indirectly dated. The first attempt at a quantitative estimate of paleodischarge contributed interesting results, but improvements are badly needed. This can be realised by increasing the number of test areas to verify the results obtained so far.

6 ACKNOWLEDGEMENTS

The authors wish to express their appreciation for the financial assistance received from the Polish Academy of Sciences (Research project MR I/25) thanks to the activities of Professor L.STARKEL, and from the A.Mickiewicz University. Comments and suggestions given by Professors MOJSKI, ROTNICKI, STARKEL and TOBOLSKI are acknowledged, as well as technical assistance for drawings and laboratory by Mgr L.KASPRZAK, Mgr B.KULUS, Mgr M.MALINOWSKA and Mgr G.PINCZAK. For the translation of the paper into English we are grateful to Dr. Z.NADSTOGA.

7 LIST OF SYMBOLS

B_I - index of braiding after Brice; Lb - the length of bars; Lc - the length of channel; R_c - radius of curvature; W - width; D - depth; F - channel width to channel depth; A - channel cross-section; Q - discharge; P - thickness of sandy layers; M - thickness of silty layers; V - velocity; H_R - hydraulic radius; S - slope; n - roughness coefficient after Strickler; d - d = d_{50}; grain size; L - meander wavelength; A_m - amplitude; h - meander height; λ - meander width; τ_c - critical shear stress; τ_o - shear stress; γ_s - specific weight of sediment; γ - specific weight of water; Mz - mean size; δ - standard deviation; Ω - stream power after Bagnold; g - acceleration due to gravity

8 REFERENCES

ACKERS, P. & F.G.CHARLTON 1970. Meander geometry from varying flows. Journ. of Hydrol., vol. XI, no. 3, pp. 230-252.
ALLEN, J.R.L. 1964. Studies in fluviatile sedimentation: Six cyclothems from the Lower Old Red Sandstone, Anglo-Welsh Basin. Sedimentology, vol. 3, pp. 163-198.
ALLEN, J.R.L. 1968. Current ripples, their relation to patterns of water and sediment motion. North Holland Publ., Amsterdam.
ALLEN, J.R.L. 1970. Studies in fluviatile sedimentation a comparison of finning-upward cyclothems, with special reference to coarse-member composi-

tion and interpretation. Journ. Sed. Petrol., vol. 40, pp. 298-323.

ANTCZAK, B. 1978. Zróżnicowanie facjalne osadów meandrowych łach wałowych Warty na północ od Śremu (Sum: Facial variation of sediments from the Warta point bar ridges north of Śrem). Bad. Fizjogr. Pol. Zach., T. XXXI, seria A, Geogr. Fiz., pp. 7-31.

ANTCZAK, B. 1985. Transformacja układu koryta i zanik bifurkacji Warty w pradolinie Warszawsko-Berlińskij i południowej części przełomu poznańskiego podczas późnego Vistulianu (Conversion of the channel pattern and cessation of the Warta river bifurcation in the Warsaw-Berlin Pradolina and the southern section of the Poznań gap during the Late Vistulian). Ph. D. Thesis. A.Mickiewicz Univ., Poznań.

ANTCZAK, B. & S. KOZARSKI 1981. Jaszkowo – Older generation of paleomeanders and point bar ridges. Symposium 'Paleohydrology of the temperate zone', Poznań, Poland'81, Guide-book of Excursions, pp. 29-30.

BAGNOLD, R.A. 1966. An approach to the sediment transport problem from general physics. U.S. Geol. Surv. Prof. Pap., 422-J.

BAKER, V.R. 1983. Paleoflood hydrologic analysis from slack-water deposits. Quaternary Studies in Poland, 4, pp. 19-26.

BORÓWKO-DŁUŻAKOWA, Z. 1969. Palynological investigations of late glacial and Holocene deposits at Konin. Geographia Polonica, 17, pp. 267-281.

BORZYSZKOWSKI, J. 1971. Rozwój i zanik swobodnego meandru Warty koło Tworzykowa (Sum: Development and decay of the Warta free meander near Tworzykowo). Bad. Fizjogr. Pol. Zach., t. XXIV, serie A, Geogr. Fiz., pp. 11-46.

BRICE, J.C. 1964. Channel pattern and terrace of the Loup River in Nebraska. US Geol. Surv. Prof. Pap., 422 - D.

BRUNSDEN, D. & J.B.THORNES 1979. Landscape sensitivity and change. Transactions, Institute of British Geographers, New Series, 4, pp. 463-484.

CHOW, V.T. 1959. Open-channel hydraulics. McGraw-Hill, New York.

CHURCH, M. 1972. Baffin Island sandurs; a study of Arctic fluvial processes. Geol. Surv. Can. Bull., 216, p. 208.

DĄBKOWSKI, L., J.SKIBIŃSKI & A.ŻBIKOWSKI 1982. Hydrauliczne podstawy projektów wodnomelioracyjnych. PWRiL, Warszawa.

FOLK, R.L. & W.C.WARD 1957. Brazos River bar: a study of the significance of grain-size parameters. Journ. Sed. Petrol., 27, pp. 3-26.

FRAZIER, D.E. & A.OSANIK 1961. Point-bar deposits, Old River Locksite, Louisiana. Trans. Gulf Coast Assoc. Geol. Soc., 11, pp. 121-137.

FRENZEL, B. 1980. Klima der Letzten Eiszeit und der und Nacheiszeit in Europa. Veröffentlichungen der Joachim Jungius-Gesellschaft für Wissenschaften im Hamburg, 44, pp. 9-46.

FRENZEL, B. 1983. On the Central-European water budget during the last 15 000 years. Quaternary Studies in Poland, 4, pp. 45-59.

FRIEDMAN, G.M. 1962. On sorting, sorting coefficients, and the lognormality of the grain-size distribution of sandstones. Journ. Geol., 70, pp. 737-753.

GARDINER, V. & R.DACKOMBE 1983. Geomorphological field manual. George Allen and Unwin, London.

GONERA, P. 1984. Zmiany geometrii koryt meandrowych Warty na tle wahań klimatycznych w późnym Vistulianie i w holocenie (Changes in geometry of the Warta meandering channels against climatic fluctuations during the Late Vistulian and Holocene). Ph. D. Thesis. A.Mickiewicz Univ., Poznań.

GOUDIE, A. 1983. Environmental change. Clarendon Press, Oxford.

GRAF, W.H. 1971. Hydraulic of sediment transport. McGraw-Hill, New York.

GRAF, W.L. 1983. Flood-related channel changes in an arid-region river. Earth Surface Processes and Landforms, 8, pp. 125-139.

GREŃ, J. 1974. Statystyka matematyczna, modele i zadania. PWN, Warszawa.

GRZEGORCZYK, M. 1970. Metody przedstawiania uziarnienia osadów. Pozn. Tow. Przyj. Nauk, Prace Kom. Geogr.-Geol., T. X, z. 2, pp. 1-74.

HANNEMANN, M. 1961. Nue Beobachtungen zur Entestehung und Entwicklung des Berliner Ustromtals zwischen Fürstenwalde (Spree) und Fürstenberg (Oder). Geologie, 10, heft 4/5, Berlin, pp. 418-434.

HARMS, J.C. & R.K.FAHNESTOCK 1965. Stratification, bed forms and flow phenomena (with example from the Rio Grande). In: G.V.MIDDLETON (ed.), Primary sedimentary structures and their hydrodynamic interpretation. Soc. Econ. Paleontol. Mineral. Spec. Publ., 12, pp. 84-115.

HICKEN, E.J. 1981. River channel changes, 2nd Conf. Fluvial Sedimentology, The University of Keele, UK, M.S.

HOWARD, A.D. 1980. Threshold in river regimes. In: D.R.COATES & J.D.VITEK (eds), Thresholds in geomorphology, George Allen and Unwin, London, pp. 227-258.

KANIECKI, A. 1976. Dynamika rzeki w świetle osadów trzech wybranych odcinków Prosny (Sum: River dynamics in the light of deposits of three chosen sections of Prosna River). PTPN, Wydz. Mat. - Przyr., Prace Kom. Geogr. - Geol., t. VII, PWN Warszawa, Poznań.

KLIMEK, K. & L.STARKEL 1981. Some Paleohydrological reconstructions. In: L.STARKEL (ed.), The evolution of the Wisłoka Valley near Dębica during the Late Glacial and Holocene. Folia Quaternaria, 53, pp. 78-83.

KOUTANIEMI, L. & R.RONKAINEN 1983. Paleocurrents from 5 000 and 1 600-1 500 BP in the main rivers of the Oulanka basin, North-Eastern Finland. Quaternary Studies in Poland, 4, ppl. 145-159.

KOWALSKA, A. 1957. Die Korrelation des Glazialen Relief mit dem diluvialen Untergrund des westpolnischen Tiefebene. INQA V Congr. Intern. Résumes des communications, Madrid-Barcelona, P. 98.

KOZARSKI, S. 1974a. Późnoglacjalne i holoceńskie zmiany w układzie koryt rzecznych niżowej części dorzecza Odry (Late Glacial and Holocene channel pattern changes in the lowland part of the Odra River drainage basin). Sympozjum Krajowe 'Rozwój den dolinnych...', Wrocław - Poznań 1984, pp. 17-19.

KOZARSKI, S. 1974b. Stanowisko Jaszkowo koło Śremu. Migracje koryta Warty na południe od Poznania w późnym glacjale i holocenie - Generacje meandrów (Site Jaszkowo near Śrem. Migrations of the Warta River channel to the south of Poznań during Late Glacial and Holocene – Generations of meanders). Sympozjum Krajowe 'Rozwój den dolinnych...', Wrocław – Poznań, 1974, pp. 46-49.

KOZARSKI, S. 1976. Air-photo Interpretation in the Geomorphological Survey of Valley Floors in the North Polish Plain. Ceskoslov. Ak. Ved. Geogr. Ust. Bruno, Stud. Geogr., 55, pp. 79-81.

KOZARSKI, S. 1981a. River channel changes in the Warta valley to the south of Poznań. Symposium 'Paleohydrology of the temperate zone', Poznań, Poland'81, Guide-Book of Excursions, pp. 6-23.

KOZARSKI, S. 1981b. Stratygrafia i chronologia Vistulanu Niziny Wielkopolskiej (Sum: Vistulian stratigraphy and chronology of the Great Poland Lowland). PAN Oddz. w. Poznaniu, seria: Geografia, 6, PWN Warszawa - Poznań.

KOZARSKI, S. 1981c. Zbrudzewo – Generations of paleomeanders. Symposium 'Paleohydrology of the temperate zone', Poznań, Polant'81, Guide-book of Excursions, pp. 41-43.

KOZARSKI, S. 1981d. Czmoniec – Younger generation of paleomeanders. Symposium 'Paleohydrology of the temperate zone', Poznań, Poland'81, Guide-book of Excursions, pp. 50-51.

KOZARSKI, S. 1983a. River channel changes in the middle reach of the Warta Valley, Great Poland Lowland. Quaternary Studies in Poland, 4, pp. 159-169.

KOZARSKI, S. 1983b. River channel adjustment to climatic change in west central Poland. In: K.J. GREGORY (ed.), Background to paleohydrology. A perspective. John Wiley and Sons, Chichester, pp. 356-370.

KOZARSKI, S. 1983c. The Holocene generation of paleomeanders in the Warta River Valley, Great Polish Lowland. Geologisches Jahrbuch, A 71, Hannover, pp. 109-118.

KOZARSKI, S., B.ANTCZAK & P.GONERA 1984. Przemiany dna doliny środkowej Warty podczas późnego Vistulianu i Holocenu (Changes in the middle reach of the Warta valley floor during the Late Vistulian and Holocene). Research project MR I/25, final report, p. 79.

KOZARSKI, S. & K.ROTNICKI 1977. Valley floors and changes of river channel patterns in the North Polish Plain during the Late Würm and Holocene. Quaestiones Geographicae, 4, pp. 51-93.

KOZARSKI, S. & K.ROTNICKI 1978. Problemy późnowürmskiego i holoceńskiego rozwoju den dolinnych na niżu polskim (Sum: Problems concerning the development of valley floors during Late-Würm and Holocene in the Polish Lowland). Pozn. Tow. Przyj. Nauk, Prace Kom. Geogr. - Geol., t. 19, PWN Warszawa-Poznań.

KOZARSKI, S. & K.ROTNICKI 1983. Changes of river channel patterns and the mechanism of valley-floor construction in the North-Polish Plain during the Late Weichsel and Holocene. In: D.J.BRIGGS & R.S.WATERS (eds), Studies in Quaternary Geomorphology, pp. 31-48.

KRYGOWSKI, B. 1961. Geografia fizyczna Niziny Wielkopolskiej (Sum: Physical geography of Great Poland Lowland). Pozn. Tow. Przyj. Nauk, Wydz. Mat- Fiz. Kom. Fizjogr., PAN.

LAMB, H.H. 1977. Climate: present, past and future. Volume 2: Climatic history and the future. Methuen and Co. Ltd. London, Barnes and Noble Books, New York.

LEOPOLD, L.B. & T.MADDOCK, Jr. 1953. The hydraulic geometry of stream channels and some physiographic implications. US Geol. Surv. Prof. Paper, 252.

LEOPOLD, L.B., M.G.WOLMAN & J.P.MILLER 1964. Fluvial processes in geomorphology. W.H.Freeman and Co., San Francisco.

LEWIN, J. 1977. Channel pattern change. In: K.J.GREGORY (ed.), River channel changes. J. Wiley and Sons, pp. 167-184.

LIMBREY, S. 1983. Archaeology and paleohydrology. In: K.J.GREGORY (ed.), Background to paleohydrology. A perspective. J.Wiley and Sons, Chichester, pp. 189-208.

MAIZELS, J.K. 1983. Paleovelocity and paleodischarge determination of coarse gravel deposits. In: K.J. GREGORY (ed.), Background to paleohydrology. A perspective. J.Wiley and Sons, Chichester, pp. 101-130.

MANGERUD, J., S.T.ANDERSON, B.E.BERGLUND & J.J.BONNER 1974. Quaternary stratigraphy of Norden, a proposal for terminology and classification. Boreas, v. 3, no. 3, pp. 109-128.

MCGOWERN, J.H. & L.E.GARNER 1970. Physiographic features and stratification types of coarse-grained point bars, modern and ancient examples. Sedimentology, 14, pp. 77-111.

MOODY-STUART, M. 1966. High and low sinuosity stream deposits with examples from the Devonian of Sptsbergen, Journ. Sed. Petrol., 36, pp. 1102-1117.

OKUNIEWSKA, I. & K.TOBOLSKI 1981. Preliminary results of paleobotanical investigations of paleomeander fill at Mechlin. Symposium 'Paleohydrology of the temperate zone', Poznań, Poland'81,

Guide-book of Excursions, pp. 39-40.

OKUNIEWSKA, I. & K.TOBOLSKI 1982. Wstępne wyniki badań paleobotanicznych z dwóch paleomeandrów Warty koło Poznania (Sum: Preliminary results from paleobotanical studies on deposits from two paleomeanders in the Warta valley near Poznań). Bad. Fizjogr. Pol. Zach., 34, seria A, Geogr. Fiz., pp. 149-160.

PARK, C.C. 1978. Stream channel and flow relationships – recent techniques illustrated by studies on some Dartmoor streams. Field Studies, 4, pp. 729-740.

PASSEGA, R. 1964. Grain-size representation by CM patterns as a geological tool. Journ. Sed. Petrol., 34, pp. 830-847.

PICKARD, M.D. & L.R.J.HIGH 1973. Sedimentary structures of emphemeral streams. Elsevier Scien. Publ., Amsterdam.

PICKUP, G. & W.A.RIEGER 1979. A conceptual model of the relationship between channel characteristics and discharge. Earth Surface Processes, 4, pp. 37-42.

RAY PULLAK, K. 1976. Structure and sedimentological history of the overbank deposits of a Mississippi River point bar. Journ. Sed. Petrol., v. 46, no. 4, pp. 788-801.

ROSE, J., C.TURNER, G.R.COOPE & M.D.BRYAN 1980. Channel changes in a lowland river catchment over the last 13 000 years. In: R.A.CULLINGFORD, D.A.DAVIDSON & J. LEWIN(eds), Timescales in geomorphology. J.Wiley and Sons, Chichester, pp. 159-175.

ROTNICKI, K. 1983. Modelling past discharges of meandering rivers. In: K.J.GREGORY (ed.), Background to paleohydrology. A perspective. J.Wiley and Sons, Chichester, pp. 321-346.

SCHEIDEGGER, A.E. 1974. Geomorfologia teoretyczna. PWN Warszawa.

SCHUMM, S.A. 1968. River adjustment to altered hydrologic regimen-Murrumbidge River and paleochannels, Australia. US Geol. Surv. Prof. Paper, 598.

SCHUMM, S.A. 1977. The fluvial system. J.Wiley and Sons, p. 338.

SCHUMM, S.A. 1979. Geomorphic thresholds: the concept and its applications. Trans. Inst. Brit. Geogr., New Series, 4, no. 4, pp. 483-515.

SCOTT, K.M. 1978. Effects of permafrost on stream channel behaviour in Arctic Alaska. US Geol. Surv. Prof. Paper, 1068.

SELLEY, R.C. 1976. An introduction to sedimentology. Academic Press, London, New York, San Francisco.

SLY, P.G., R.L.THOMAS & B.R.PELLETIER 1983. Interpretation of moment measures derived from waterlain sediments. Sedimentology, v. 30 no. 2, pp. 219-233.

ŚRODOŃ, A. 1972. Roślinność Polski w czwartorzędzie. In: W.SZAFER & K.ZARZYCKI (eds), Szata roślinna Polski, t. 1, PWN Warszawa, pp. 527-569.

STARKEL, L. & J.B.THORNES 1981. Paleohydrology of River Basins. British Geomorphological Research Group, Technical Bulletin, no. 28.

TOBOLSKI, K. 1966. Późnoglacjalna i holoceńska historia roślinności na obszarze wydmowym w dolinie środkowej Prosny (Sum: Late Glacial and Holocene history of vegetation in the dune area of the Middle Prosna Valley). Pozn. Tow. Przyj. Nauk, Wydz. Mat. - Przyr., Prace Kom. Biol., t. 22, z. 1.

UNRUG, R. 1959. Spostrzeżenia nad sedymentacja warstw lgockich. Rocz. Pol. Tow. Geol., 20, pp 197-225.

WILLIAMS, P.F. & B.R.RUST 1969. The sedimentology of a braided river. Journ. Sed. Petrol., 39, pp. 649-679.

Lake, Mire and River Environments, Lang & Schlüchter (eds)
© 1988 Balkema, Rotterdam. ISBN 90 6191 849 9

The problem of the hydrological interpretation of palaeochannel pattern

J.Rotnicka
Institute of Physical Geography, Adam Mickiewicz University, Poznań, Poland

K.Rotnicki
Quaternary Research Institute, Adam Mickiewicz University, Poznań, Poland

ABSTRACT

The channel pattern is often used in palaeohydrology as an indicator of the hydrologic regimen. The term 'hydrologic regimen', however, is variously understood, which results in the ambiguity of using channel pattern as an indicator of past hydrological conditions. This article presents the results of testing the hypothesis that a river channel pattern is a result of discharge variability; a meandering channel – by this hypothesis – is the effect of a small variability of discharges, while a braided channel results from their great variability. Research on discharge variability for 28 reaches of Polish and Yugoslav rivers with specific channel patterns leads to a rejection of this hypothesis for rivers of the temperate climate.

1 INTRODUCTION

The river channel pattern and its changes are a subject of particular interest in the studies of the palaeohydrology of river valleys, or palaeopotamology. The studies were introduced about 20 years ago by DURY (1964, 1965), SCHUMM (1965, 1968, 1969) and FALKOWSKI (1965, 1967, 1971). The river channel pattern became the basis for the investigations of hydrologic conditions prevailing in the river valleys in the past periods of the Quaternary.

However, only channels from the Holocene and the last cold period, especially its final phase, have been preserved in a subfossil form, i.e. a form which is relatively easy to detect by means of air photography, exact hypsometric maps and shallow geological soundings. For this reason, in a majority of works on palaeopotamology, the analysis of channel patterns covers the period of the last 15,000-18,000 years. The determination of the channel pattern, the shape and size of a channel for a given period and an analysis of its transformations in the discussed time interval, have become a necessary starting point in the palaeohydrologic analysis of river valley floors.

2 CHANNEL PATTERN AS A TOOL OF PALAEOHYDROLOGIC ANALYSIS

Studies by DURY (1964a, b, 1965), SCHUMM (1965, 1968, 1969) and in Poland by FALKOWSKI (1965, 1967, 1971), inspired a growth of interest in channel size, shape and pattern. The authors came to the conclusion that the differences in size, shape and pattern of channels of various ages resulted from the influence of climatic changes on the hydrologic regimen. For the last few hundred years, the shape and pattern of a channel have also been increasingly affected by the activity of man.

Because of these conclusions, numerous publications have appeared since the mid-1960s, both general and specific, which analysed changes in both channel patterns and such geometric features of meandering channels as their width and depth, curvature radii, meander wave lengths, etc., for the last 15,000 years (FALKOWSKI 1971, 1975, 1980, 1982; KLIMEK & STARKEL 1974; KOZARSKI 1974a, 1974b, 1981, 1983a, 1983b; MYCIELSKA-DOWGIAŁŁO 1972, 1977; ROTNICKI 1974; KOZARSKI & ROTNICKI 1977, 1983; DURY 1976, 1977; ALLEN 1977; FROELICH et al. 1977; GREGORY 1977; HICKIN 1977; HITCHCOCK 1977; KNIGHTON 1977; LEWIN 1977, 1978, 1983; PARK 1977; THORNES 1977; ROSE et al. 1980; SZUMAŃSKI 1972, 1982; STARKEL 1981, 1983; HEINE 1982; TOMCZAK 1982; WIŚNIEWSKI 1982, STARKEL et al. 1982; ANDRZEJEWSKI 1984). In addition, two books edited by GREGORY (1977, 1983) have been published and aim at a more general and broader view of various aspects of channel changes, their causes and palaeohydrologic significance. Finally, on the basis of channel geometry and features of channel deposits, attempts have been made to determine the volume of past river discharge (DURY 1964a, 1964b, 1965; SCHUMM 1968, 1969; MAIZELS 1983a, 1983b; ROTNICKI 1983a, 1983b).

The conclusions drawn from the analysis of changes of geometric features of meandering channels and changes in the channel pattern can be reduced to three fundamental points:

1) The transformation of the braided channel patterns into meandering patterns at the turn of the last cold period and the Holocene resulted from changes of the hydrologic regimen of a river and its basin under the influence of a changing climate. The high frequency of this phenomenon may suggest that braided channels are characteristic of the climatic and hydrological conditions of the cold zone, while meandering channels are connected with the conditions of the humid temperate zone.

2) Changes of climate and vegetation during the Holocene were followed by changes in the channel-forming discharge and sediment yield; these, in turn, determined the channel pattern and erosive or accumulative tendencies of a river, which is what STARKEL (1983) tried to show applying the methodological approach of SCHUMM (1969) to the problem of river metamorphosis.

3) The tendency towards the transformation of channel patterns from meandering into braided patterns during the last 300 years is connected with artificial deforestation, i.e. with the activity of man who has changed woodland into farmland (FALKOWSKI 1965, 1967, 1971).

3 THE PROBLEM – A SURVEY OF PUBLISHED OPINION

In the light of such a widespread use of the channel pattern as a tool of hydrological analysis in palaeopotamology, the following problem appears: 1) if the analysis of the channel pattern and geometry is so important as a research tool and what kind of hydrological information do they contain? and, 2) does the information concern the hydrologic regimen? If so, how is the latter understood? What is meant here is data that can be deciphered and expressed in terms of quantitative measures used in hydrology.

In hydrology, the term 'hydrologic regimen' has not got an unambiguous meaning. Usually, however, it is understood as the annual discharge variability determined by the source and kinds of water supply within the framework of specific physiographic conditions of the drainage basin (PARDE 1949; KUZIN 1955, 1960; DĘBSKI 1970; DYNOWSKA 1972; ROTNICKA 1976, 1977). It would be useless, however, to look in hydrological literature for any connection or its lack between a definite type of channel pattern and a definite type of hydrologic regimen. In palaeohydrology, we can distinguish two approaches to the cause-response relation between the hydrologic regimen and the type of channel pattern. Depending on the approach, we get a different scope of palaeohydrologic interpretation of the preserved past channel patterns.

The different approaches result from different conceptions of both the hydrologic regimen and the main causes of the formation of a channel of definite geometrical features determining its shape and pattern. Frequently enough, the authors do not define the hydrologic regimen, so it is possible that what they differ in is not the conception of the regimen, but of its basic features and their relative prominence. The two approaches can be summarized as follows:

1. One approach treats the channel as a result of the influence of precipitation and temperature on runoff and sediment yield. In other words, the channel pattern is treated as a linkage between climate and drainage basin properties. Annual discharge variability to define the hydrologic regimen and as the factor determining the channel pattern is of minor importance, and sometimes is ignored altogether.

2. In the other palaeohydrological approach, the hydrologic regimen is understood as modern hydrology usually understands it, viz. as the variability of river discharge in an annual cycle determined by the source and kind of water supply against the background of the physiographic conditions of the drainage basin. The channel pattern is thus a function of discharge variability. This approach was formulated by FALKOWSKI (1971, 1975, 1980). Both approaches will be discussed below.

3.1 Channel pattern as a result of the influence of the climate on the runoff and sediment yield

In the analysis of the palaeochannel pattern, we usually encounter the first approach, as given by DURY (1964a, 1964b, 1965) and SCHUMM (1965, 1968, 1969). The matter was particularly exhaustively explored by SCHUMM when he formulated the problem of 'river adjustment to altered hydrologic regimen' (1968) and of 'river metamorphosis' (1969). This eminent scholar never stated his understanding of the hydrologic regimen explicitly, but it follows from his work that he treated it as the whole of hydrologic relations of the drainage basin conditioned by such independent variables as precipitation and temperature. These, in turn, influence such dependent variables as evaporation, runoff/precipitation ratio, mean annual discharge, mean annual flood, sediment yield, and type of sediment load moved through the channel. The changes of the regimen understood in this way, in turn, bring about changes in the channel size, shape, sinuosity, and gradient, and the pattern of alluvial channels. SCHUMM (1968, 1969) strongly emphasizes the role of the type of transported material along with changes in mean annual discharge and mean annual flood, in the formation of the channel pattern (SCHUMM 1960, 1963, 1968, 1969). He gives the relationships between the width/depth ratio, sinuosity, and meander wave length, and the kind of material found in the channel perimeter. He expresses these relations by means of several equations, which can be given a more general form, namely:

$$F, P = f'(M) \tag{1}$$

$$L = f'(M, Q) \qquad (2)$$

where F is the width/depth ratio, P is sinuosity, M is the silt-clay percentage in the channel perimeter, Q is discharge and f' is a derivative. SCHUMM (1968, 1969) shows how intricate and many-sided the relations and linkages are between the discussed variables and the channel shape and pattern. This scholar's works also show how varying the responses of channels and river systems to climatic changes can be. For instance, an increase in precipitation can produce different effects in a channel. It depends on the initial state of such climatic variables as precipitation and temperature, and on the initial physiographic conditions of the drainage basin. That is why SCHUMM does not attempt to relate a definite channel pattern either to definite climatic conditions, or to a definite hydrologic regimen.

LEWIN (1977) considers the causes of channel pattern changes in more general terms. He distinguishes two kinds of such changes: 1) autogenic, 'inherent in the river regime', comprising migration, cut-offs, avulsion, etc., and 2) allogenic, which appear in response to the changes in the climate, discharge and sediment load. LEWIN (1977) does not relate specific channel patterns to specific climatic and hydrologic conditions, either. He follows SCHUMM (1969, 1971, 1974) in pointing out that the complete transformation of river morphology, called metamorphosis by SCHUMM (1969), need not necessarily be allogenic. Lewin states that '. . . the complication lies in the apparent existence of abrupt thresholds between pattern states, with sudden changes occurring when critical limits are exceeded (SCHUMM 1974). Hence, a small change in slope may lead to a large change in channel pattern. . .' (LEWIN 1977: 179).

A similar position is occupied by KNOX (1982). He is far from attributing the same reaction to specific climatic changes to all the rivers of a given area. On the contrary, he is of the opinion that all generalization of responses of river systems to climate change is difficult to formulate.

It follows from the above survey of opinions that definite changes in the hydrologic regimen can produce different effects in different rivers and, hence, that a definite channel pattern cannot be related to definite climatic conditions and hydrologic regimen.

3.2 Channel pattern as a result of climatically controlled discharge variability

The other approach to understand the relationship between channel pattern and hydrologic regimen, understood here as a spectrum of discharges, is particularly well expressed in FALKOWSKI (1971, 1972, 1975, 1980). According to his hypothesis, channel pattern changes – whether under the influence of climatic changes or man's activity – occur in a 'fluviodynamic cycle' (1975). The cycle covers channel pattern changes from a braided to a meandering river and vice

versa. Between these two extreme types a transitional phase with 'transitional patterns' occurs. A river can respond to climatic changes and man's activity with a full, undisturbed cycle of channel development and channel pattern changes only when its channel is mature and free. According to FALKOWSKI (1971, 1972, 1975), this is the case when: 1) The bottom of alluvial deposits on the valley floor lies below the range of influence of channel processes during high discharges; 2) the valley floor is wide enough for free meandering; 3) there are no obstacles on the valley floor in the form of alluvial fans, dunes, etc. Generally, the lack of these conditions was considered by LEWIN & BRINDLE (1977) to produce confined meanders.

According to FALKOWSKI (1971, 1972, 1975), there is a specific channel pattern corresponding to a specific hydrologic regimen, understood, to use LEWIN's (1977) apt expression, as a spectrum of discharges. Thus, a hydrologic regimen characterized by very large discharge amplitudes produces a braided channel pattern, while a 'mild' regimen with maximally equalized discharges and mild and small floods leads to a meandering channel pattern. If we have a free channel development, a change of the hydrologic regimen induces a change of the channel from braided to meandering or vice versa, with different transitional phases (Fig. 1). Note that in the hypothesis presented in Figure 1, on the figure in A shows a situation resembling a real one, while the models of the remaining situations are only theoretical (Figs. 1B-E). A confined river cannot change its channel pattern in spite of a change of the hydrologic regimen following FALKOWSKI's concept.

STARKEL (1977) and SZUMAŃSKI (1972) also expressed the opinion that the braided channel pattern was connected with large and the meandering with small discharge amplitudes. KOZARSKI & ROTNICKI (1977), in turn, stated that a channel pattern resulted from the volume and the degree of irregularity of discharges, among other things, but they did not elaborate on how the channel pattern was connected with this feature of the river. Recently, STARKEL (1983) constructed a model of changes in aggradation and erosive tendencies of rivers and of channel pattern changes as a response to change of the hydrologic regimen of Central European rivers in the last 15,000 years. This may be interpreted as his endorsing the view of a uniform response of the river channel pattern of a given climatic zone to changes of the climate and hydrologic regimen. However, we do not know how the hydrologic regimen is understood in this model.

Let us use the systems analysis (CHORLEY & KENNEDY 1971) to understand the two different approaches to the relation between channel size and pattern and the hydrologic regimen. In such an approach, the channel pattern is the result of the river process-response system. In this process-response system, the main input is precipitation and the output is evaporation, water discharge, and sediment transport.

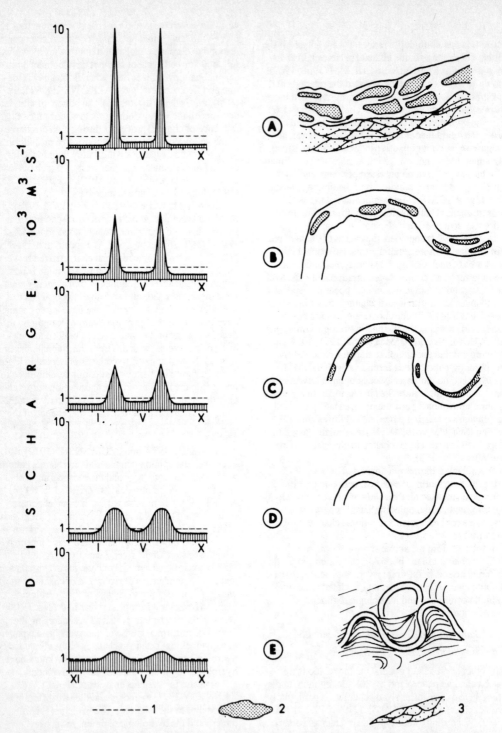

Figure 1. Diagram of changes in the hydrological regimen of rivers and corresponding phases in channel pattern changes according to FALKOWSKI (1980). Mean annual discharge as for the middle Vistula. Dashed line represents mean annual discharge.

And the main regulators of the river channel are: temperature, vegetation, infiltration capacity, and erodability. Erodability should be treated as a complex effect of an interaction of flow velocity, particle size in channel perimeter, and the density and compactness of the vegetation, especially its root system in the channel bed and banks.

It can be stated that in SCHUMM's approach (1968, 1969) the mentioned inputs, output and regulators of the cascade system constitute the notion of the hydrologic regiment.

When accepting the complexity of the river process-response system, it is obvious that: 1) change of various elements of the hydrologic regimen can lead to similar changes of the channel shape and pattern, and also the other way around, 2) a change of the same elements of the hydrologic regimen need not induce the same changes of the channel pattern and shape in various rivers each time it occurs, because the values of the system regulators determining certain thresholds are a specific feature of a given drainage basin. This is the reason why a specific type of channel shape and pattern cannot be related to a definite feature of the hydrologic regimen. Therefore, the channel pattern is a manifestation in SCHUMM's approach (1965, 1968, 1969), of the hydrologic regimen systematically understood as the hydrologic regimen of a drainage basin, which is nothing else but numerical values of inputs, outputs and regulators of the cascade system of the hydrologic cycle of the drainage basin. In turn, in the other approach by FALKOWSKI (1971, 1975, 1980), the channel pattern is a reflection of the variability of a river's discharges. In this understanding, the regimen amounts to the visible spectrum of the hydrologic regimen of the river basin.

In conclusion, the 'channel pattern' is used in palaeohydrology as the criterion for defining the hydrologic regimen in two meanings: as the hydrologic regimen of a river basin, or as the hydrologic regimen of a river, i.e. the variability of its discharges.

4 THE TESTED HYPOTHESIS

The relationship between the channel pattern and the regimen understood as a spectrum of discharges has not been studied yet, either by hydrology or by dynamic geomorphology. Hence, it is still a mere hypothesis. The observation of modern channel processes seems to indicate that the channel pattern is a function of flow velocity and erodability of the channel bed and banks (LEOPOLD, WOLMAN & MILLER 1964; BRICE 1964; LANE 1957; FRIEDKIN 1945; LEOPOLD & WOLMAN 1957). This also follows indirectly from formulae (1) and (2).

This dilemma between the hypothesis used sometimes in palaeohydrological research as the criterion for determining the discharge variability of past rivers and the observation of modern channel processes was the

stimulus for the present authors to examine the following problem: are there hydrological determinants of river channel patterns? The examination also aimed at applying in palaeohydrological analysis only those relations and regularities which are proved in the field.

If we were to assume, as some scholars do, that the hydrologic regimen of a river understood as a discharge spectrum determined the channel pattern type, we could reasonably infer that discharge spectra for rivers of different channel patterns would also differ, while for rivers of the same type of channel pattern, it would be the same or very similar. In practice, the verification of this hypothesis amounts to finding numerical measures for: 1) the description and comparison of discharge spectra of rivers of different channel patterns, and 2) testing, by means of cluster analysis, whether the groups of the river reaches under examination, distinguished according to the criterion of discharge variability, coincide with the specified types of channel pattern.

5 THE OBJECTS OF RESEARCH

Twenty-eight river reaches of varying channel patterns were chosen for the study. Twenty-four of them are situated in Poland (Fig. 2). They represent the zone of the temperate transitional climate from marine to continental. For comparison, four river reaches were also chosen from the zone of warm, continental, arid temperate climate. These are three reaches of the Vardar in Macedonia with a braided channel pattern and one reach in the lower course of the Velika Morava in Serbia with a free meandering channel pattern. Discharge variability was examined for a total of nine reaches of rivers with braided channels, eight with transitional channels, and eleven with meandering channels. The channel patterns of 50% of the reaches were determined by FALKOWSKI (1971). For the rest, except one, the pattern was determined by the present authors on the basis of their own observations and the analysis of air photographs (Table 1, Fig. 2). There is a gauging station at the end of each of the reaches or within their limits. Daily water stages and discharges in these stations in the period 1950-1975 formed the basis for the analysis of discharge variability of particular river reaches.

6 RESEARCH METHODS

The criteria used to ascribe the type of braided or free meandering channel, or one with traces of confined meandering, to particular reaches, were those defined by LEOPOLD & WOLMAN (1957), LEOPOLD, WOLMAN & MILLER (1964), BRICE (1964), LEWIN & BRINDLE (1977) and FALKOWSKI (1971, 1975). The type of transitional channel was defined following FALKOWSKI (1971, 1975). It is

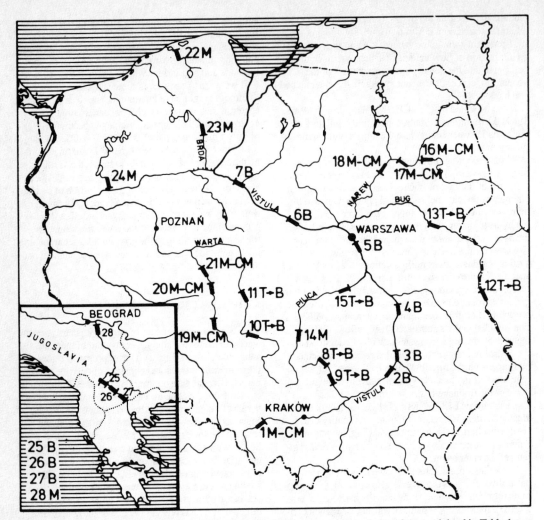

Figure 2. Location of the analysed river reaches. Numbers of reaches and gauging stations explained in Table 1. B, braided channel; T→B, transitional channel with clear traces of braiding; M, meandering channel; CM, meandering channel with traces of confinement.

characterized by smaller sinuosity than that of the meandering channel, the appearance of central bars, and a reduced width/depth ratio (Fig. 1B-C). The occurrence of central bars is the diagnostic criterion for this type of channel. In the test cases of meandering channels with traces of confined meandering analysed, the main factors producing the irregularity of the development of particular meander loops are: lithological heterogeneity with the resultant differences in the erodability of alluvial deposits of the valley floor caused precisely by the meandering of the river, as well as the presence of trees with their root systems on its banks.

In the analysis of discharge variability, the following measures were used:

1. The coefficients of characteristic monthly discharges: w_j^{max}, w_j^m, and w_j^{min}. These are relative measures characterizing the runoff distribution in an annual cycle. They make it possible to determine not only the number of periods of discharge above or below the average, but also discharge amplitude. Since they are relative measures, they allow a comparison of runoff distributions of rivers of various magnitude. The coefficients were calculated according to the following formulae:

$$w_j^{max} = \frac{\sum\limits_{i=1}^{n} Q_{ij}^{max}}{n} \cdot \frac{1}{Q_m} \tag{3}$$

Table 1. Measures of discharge variability for river reaches with varying channel patterns.

River	Name of reach cf. Fig. 2	No. of reach	Channel pattern	Measures of discharge variability from probability curve				from flow-duration curve	monthly amplitudes	
				$\bar{k}_i$	$k_{50\%}$	C_v	s	z	mean $\bar{A}_Q$	standard deviation
Vistula	Nowy Bieruń	1	M CM	76.1	68.0	0.926	0.86	10.1	2,82	1.72
	Sandomierz	2	B	33.8	28.5	0.974	0.77	6.0	2.05	1.15
	Zawichost	3	B	30.1	23.0	0.956	0.97	6.3	1.93	1.05
	Puławy	4	B	23.4	20.5	0.932	.067	5.7	1.56	0.81
	Warszawa	5	B	18.7	15.5	1.000	0.90	5.4	1.35	0.69
	Płock	6	B	–	–	–	–	5.2	1.09	0.57
	Toruń	7	B	12.8	11.5	0.691	0.77	5.0	0.98	0.50
Nida	Brzegi	8	T B	29.0	22.5	1.040	1.03	5.5	2.22	1.35
	Pińczów	9	T B	26.8	22.0	0.886	1.03	5.3	1.86	1.20
Warta	Działoszyn	10	T B	–	–	–	–	3.4	0.98	0.43
	Sieradz	11	T B	9.9	8.8	0.670	3.9	1.06	0.52	
Bug	Włodawa	12	T B	24.4	20.0	0.912	0.74	9.0	1.04	1.07
	Frankopol	13	T B	17.6	15.8	0.759	0.70	8.9	0.88	0.78
Pilica	Przedbórz	14	M	15.9	13.0	0.981	0.98	4.5	1.43	0.67
	Białobrzegi	15	T B	13.5	12.0	0.717	0.79	4.2	1.39	0.73
Narew	Strękowa Góra	16	M CM	19.3	16.0	0.953	0.88	7.8	0.94	0.88
	Piątnica Łomża	17	M CM	14.4	12.0	0.917	0.91	6.3	0.78	0.77
	Ostrołęka	.18	M CM	10.5	10.0	0.700	0.57	5.2	0.70	0.64
Prosna	Mirków	19	M CM	47.2	35.0	1.000	0.97	5.7	1.87	1.19
	Piwonice	20	M CM	24.9	19.5	1.026	1.25	7.3	1.47	1.02
	Bogusław	21	M CM	26.3	22.0	0.795	0.86	7.3	1.37	0.93
Łupawa	Smołdzino	22	M	4.1	3.3	0.788	1.08	1.8	0.46	0.08
Brda	Tuchola	23	M	2.6	2.4	0.417	1.00	1.8	0.32	0.04
Drawa	Drawiny	24	M	2.7	2.6	0.288	0.40	2.2	0.24	0.07
Vardar	Skopje	25	B	19.0	16.0	1.219	1.23	–	1.30	0.58
	Titov Veles	26	B	24.8	19.5	1.205	1.10	–	1.63	0.81
	Demir Kapija	27	B	32.0	23.0	1.348	1.48	–	1.52	0.75
Velika Morava	Lubicevski Most	28	M	34.7	–	–	–	–	1.44	0.91

Explanations:
B – braided channel
M – free meandering channel
CM – meandering channel with traces of confined meanders
T B – transitional channel with distinct traces of braiding

Determination of channel pattern according to: FALKOWSKI 1971, the authors, MYCIELSKA-DOWGIAŁŁO 1972, 1977.

$$w_j^{mean} = \frac{\sum_{i=1}^{n} Q_{ij}^{mean}}{n} \frac{1}{Q_m} \qquad (4)$$

$$w_j^{min} = \frac{\sum_{i=1}^{n} Q_{ij}^{min}}{n} \frac{1}{Q_m} \qquad (5)$$

where w_j^{mean} is the coefficient of mean monthly discharge, w_j^{max} is the coefficient of mean maximum monthly

discharge, w_j^{min} is the coefficient of mean minimum monthly discharge, Q_j^{mean} is the mean monthly discharge, Q_j^{max} is the maximum monthly discharge, Q_j^{min} is the minimum monthly discharge, Q_m is the mean annual discharge, n is the number of analysed years, j denotes a given month, and i successive years. Making use of the distribution of coefficients w_j^{max}, w_j^{mean}, and w_j^{min}, one can study discharge variability along the course of a river, or analyse the similarities and differences in dis-

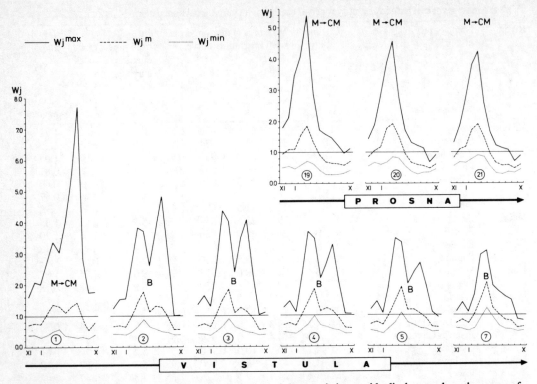

Figure 3. Changes in the annual distribution of coefficient of characteristic monthly discharges along the course of a river and the channel pattern, on the example of the Vistula and the Prosna. For the numbers and location of the reaches, see Table 1 and Figure 2.
M, free meandering channel; CM, meandering channel with traces of confinement; T, transitional channel with clear traces of braiding; B, braided channel.

charge variability between chosen points of various rivers (Fig. 3, 4).

2. The coefficient of annual discharge amplitude k_i and its probability distribution. The coefficient was calculated according to the formula:

$$k_i = \frac{Q_i^{max}}{Q_i^{min}} \qquad (6)$$

where Q_i^{max} is the maximum annual discharge, Q_i^{min} is the minimum annual discharge, and i is a given year.

The coefficient of annual discharge amplitude contains information on the range of discharge values during a year. High values of coefficient k_i indicate poor retentive capacity of the drainage basin (DYNOWSKA 1972). The strong dependence of coefficient k_i upon the water supply of a river is the reason it can change considerably from year to year. That is why the probability distribution of coefficient k_i characterizes the irregularity of extreme discharges much better than mere absolute values of k_i for particular years.

In the analysis of the probability distribution of

coefficient k_i DĘBSKI's (1954) distribution was assumed, based on decile measures: d_1, d_5, d_9. The ordinates of the probability curve are calculated from a functional equation of the form:

$$a_{p\%} = d_5 [1 + C_v \Phi(p, s)] \qquad (7)$$

where d_5 is the middle decile denoting a value of k_i corresponding to empirical probability $p = 50\%$, C_v is the variation coefficient calculated from the formula $C_v = (d_1 - d_9)/2d_5$, s is the skewness coefficient calculated from the formula $s = 2(d_1 + d_9 - 2d_5)/(d_1 - d_9)$, and Φ is a function dependent on the skewness and probability coefficients s and p. The probability curves for the occurrence of a specified k_i for rivers of varying channel patterns are illustrated in Figure 5.

3. Cumulative curves of the duration of specified and higher discharges (Fig. 6) and variability measures z calculated on the basis of these curves (Table 1). The variability measure is expressed by the formula:

$$z = \frac{Q_{30}}{Q_{330}} \qquad (8)$$

212

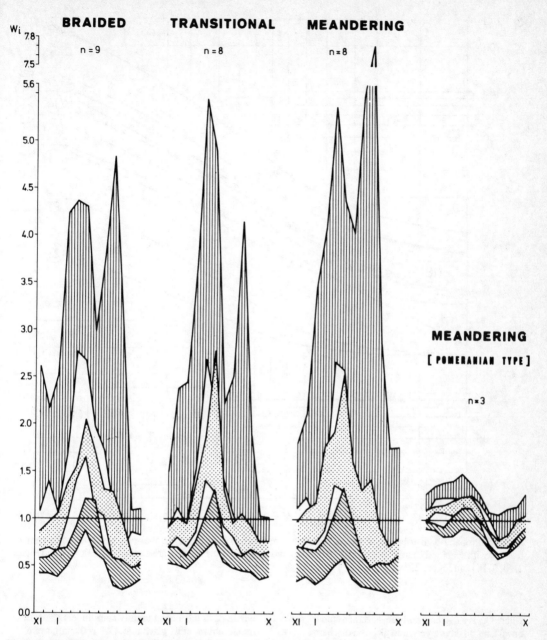

Figure 4. Variation intervals of coefficients of characteristic monthly discharges w_j^{max}, w_j^{mean}, w_j^{min} for sets of river reaches with a specified channel pattern.

Vertical lines represent variation interval of w_j^{max}, dotted area represents variation interval of w_j^{mean}, and oblique lines represent variation interval of w_j^{min}.

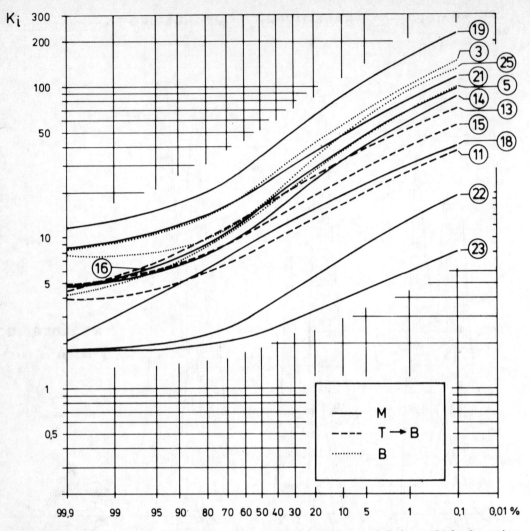

Figure 5. Probability distribution of the coefficient of irregularity of extreme annual discharges (k_i) for chosen river reaches with specified channel patterns. The number and location of the reaches are explained in Table 1 and Figure 2. M, T, B, as in Figure 2.

where Q_{30} is a discharge which, together with higher ones, lasts 30 days a year, and Q_{330} is a discharge which, together with higher ones, lasts 330 days a year.

4. The distribution of monthly amplitudes and its statistical measures (Table 1): mean monthly amplitude $\bar{A}_Q$ and standard deviation (σ). In order to compare the distribution on monthly amplitudes for rivers of different discharges, a relative measure was used; that is why the formula for the calculation of mean monthly amplitude has the form:

$$\bar{A}_Q = \sum_1^{mn} \left(\frac{Q_{ij}^{max} - Q_{ij}^{max}}{\bar{Q}} \right) \frac{1}{mn} \qquad (9)$$

where $\bar{A}_Q$ is the mean monthly amplitude, Q_{ij}^{max} is the maximum monthly discharge, Q_{ij}^{min} is the minimum monthly discharge, $\bar{Q}$ is several years' mean annual discharge, i is a month (i = 1, 2, ..., m), and j is a year (j = 1, 2, ..., n).

7 RESULTS

The analysis of the discharge variability of rivers with varying channel patterns yields the following results:
1. There is a great diversity of numerical measures of discharge variability. Their values are independent of

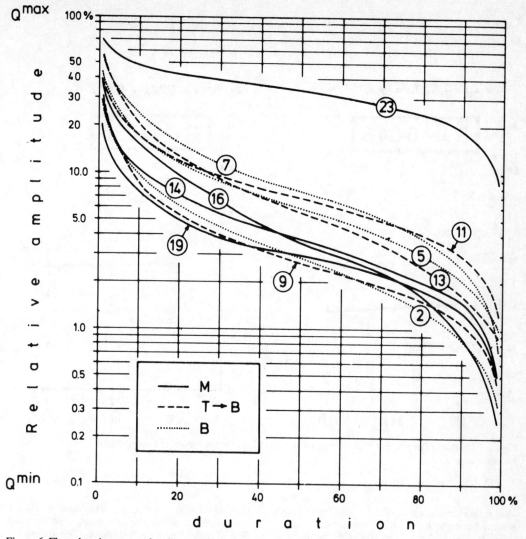

Figure 6. Flow-duration curves for chosen river reaches with a specified channel pattern expressed in relative values for the period 1951-1970; maximum discharge = 100%, minimum discharge = 0%. The numbers and location of the reaches are explained in Table 1 and Figure 2.

the channel pattern (Table 1, Fig. 7). For instance, the greatest values of the discharge irregularity coefficient $k_{50\%}$ characterizes the Vistula at Nowy Bieruń (68.0) and the Prosna near Mirków (35.0). Both rivers have meandering channels at these points, with traces of confined meandering. At the same time, the meandering Grda near Tuchola has the lowest coefficient (2.4). The remaining reaches of meandering channels, as well as the analysed braided and transitional channels, are characterized by intermediate values of coefficient $k^{50\%}$. Coefficient z for meandering rivers

ranges from 10.1 at Nowy Bieruń on the Vistula at 1.8 at Tuchola on the Brda (Fig. 7). For braided and transitional channels z assumes intermediate values (Fig. 7).

2. The coefficients of discharge variability w_j^{max}, w_j^{mean}, w_j^{min}, k_i, $k_{50\%}$, and z tend to decrease downvalley of a river (the rivers Vistula, Prosna, Bug, Narew – Fig. 3, Table 1). The decrease of discharge variability down the course of a river seems to be a rule, while the specific differentiation of this coefficient along the course of the rivers is controlled by local and possibly climatic properties of a given drainage basin. For the

215

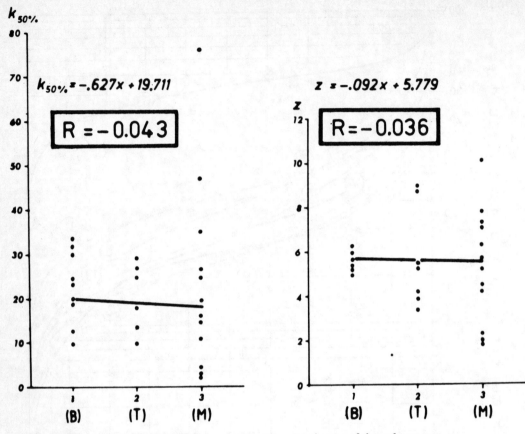

Figure 7. Relation of coefficient of discharge variability ($k_{50\%}$, z) to the type of channel pattern. r is correlation coefficient.

Vardar, for example, a river of the dry continental temperate climate, discharge variability increases downvalley. However, what is important for the problem under study is the fact that the type of channel pattern is not connected either to the values of the variability, or to its type of spatial differentiation (Table 1, Figs 3, 4, 5, 6, 7). The Vistula has a meandering channel at Nowy Bieruń, where discharge variability is greatest, while from Sandomierz to Toruń, its channel is braided, in spite of smaller, and still decreasing, variability. The Prosna has a meandering channel, although it displays greater discharge variability than the middle Vistula. These observations show that the type of channel pattern is not a result of discharge variability; in other words, the coefficients of discharge variability are not indicators of the channel pattern. It stands out clearly in the diagrams of measures w_j^{max}, w_j^{mean}, and w_j^{min} grouped according to the channel pattern (Fig. 4). When comparing these diagrams, it is hard to indicate significant differences in the discharge variability of the three

channel patterns. But what is striking is the significant difference between two types of meandering rivers, one (Fig. 4C and D) being in Pomerania. Their drainage basins are located in a young glacial area with numerous glacial kettle-holes and lakes. Their headwaters flow across lakes and their basins are forested. Hence their retention is very large. This example shows to what degree the coefficients of discharge variability are determined by regional factors within one climatic zone. Even such extreme differences in discharge variability as those in Figure 4C and D do not produce different channel patterns. In the cases under examination, meandering rivers are characterized by both very great and very small variability of their discharges. Rivers with braided and transitional channel patterns show intermediate values of the discharge variability coefficients. It should be kept in mind, however, that the analysis embraced rivers of a relatively small area. The problem requires investigation for other areas and other climatic zones.

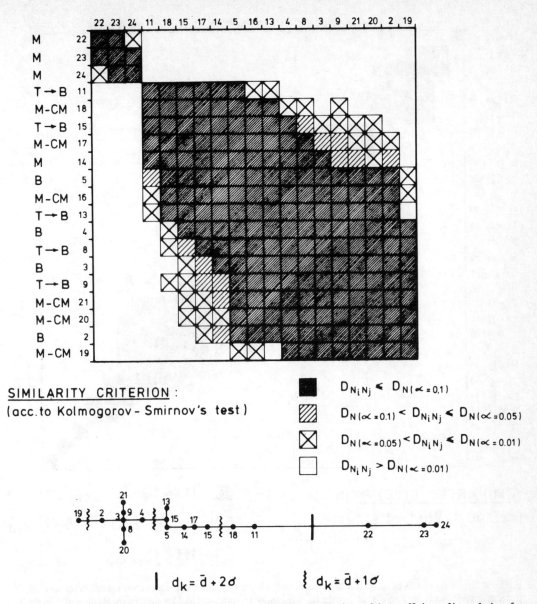

SIMILARITY CRITERION :
(acc. to Kolmogorov - Smirnov's test)

$$D_{N_i N_j} \leqslant D_{N(\alpha = 0,1)}$$

$$D_{N(\alpha = 0.1)} < D_{N_i N_j} \leqslant D_{N(\alpha = 0.05)}$$

$$D_{N(\alpha = 0.05)} < D_{N_i N_j} \leqslant D_{N(\alpha = 0.01)}$$

$$D_{N_i N_j} > D_{N(\alpha = 0.01)}$$

$$d_k = \bar{d} + 2\sigma \qquad \{ \ d_k = \bar{d} + 1\sigma$$

Figure 8. Similarity matrix and the shortest dendrite of probability distributions of the coefficient of irregularity of extreme annual discharges (k_i) for chosen river reaches with specified channel patterns from Poland. For the explanation of the numbers and location of the reaches, see Table 1 and Figure 2; M, CM, T, B, as in Figure 4.

3. The criteria of the measures of discharge irregularity k_i and $\bar{A}_Q$ do not allow the river reaches under study to be grouped by channel patterns. This conclusion follows from the KOLMOGOROV-SMIRNOV test of the significance in differences of the probability distributions of coefficient k_i and the mean monthly amplitudes $\bar{A}_Q$. On the basis of the differences in the probability distribution k_i and in mean amplitudes $\bar{A}_Q$ of all

the examined river reaches, a similarity matrix was constructed (Figs 8, 9). Through arranging the matrix of similarities among probability distributions of k_i, two typological classes of the river reaches were obtained (Fig. 8). The same classes were obtained with the dendrite method, using the criterion $d_k = \bar{d} + 2$, where d_k denotes critical distance, and $\bar{d}$ mean distance. The first group includes three Pomeranian meandering

217

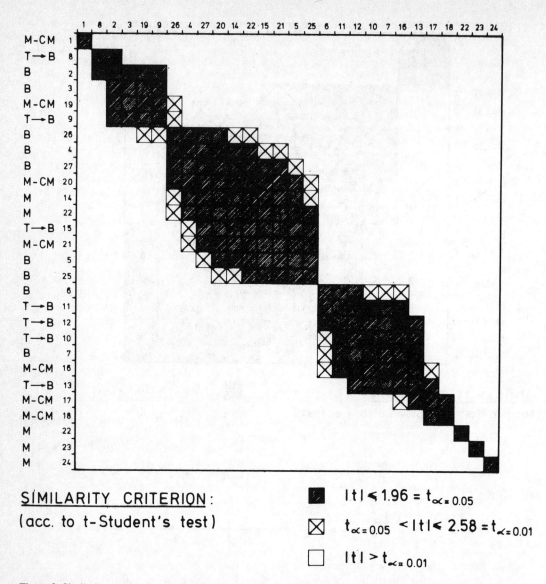

SÍMILARITY CRITERION:
(acc. to t-Student's test)

◼ $|t| \leqslant 1.96 = t_{\alpha = 0.05}$

⊠ $t_{\alpha = 0.05} < |t| \leqslant 2.58 = t_{\alpha = 0.01}$

☐ $|t| > t_{\alpha = 0.01}$

Figure 9. Similarity matrix of mean discharge amplitude (A_Q) for chosen river reaches with specified channel patterns from Poland and Yugoslavia. For the explanation of the numbers and location of the reaches, see Table 1 and Figure 2; M, CM, T, B, as in Figure 4.

rivers, and the other all the remaining ones, irrespective of their channel patterns. The application of a stricter criterion of division ($d_k = \bar{d} + 1$) to the river reaches of the second group made it possible to distinguish further three subgroups, but with very indistinct boundaries. However, they do not arrange the rivers by their channel patterns either. The matrix of similarities of mean monthly amplitudes $\bar{A}_Q$ does not group the rivers by their channel pattern either (Fig. 9). Each group dis-

tinguished in the matrix contains all the three types of channel patterns, the Pomeranian rivers form a separate group which is characterized by a lack of similarities among their constituent elements.

8 DISCUSSION OF RESULTS

The results presented above make us reject the hypothesis about the 'mildness' of the hydrological regimen and small discharge variability of meandering rivers, and about the great irregularity and discharge variability of braided rivers. As SCHUMM states (1968: 41), 'The differences between types of river channels are most marked in rivers of subhumid, semi-arid, and arid regions. The major perennial streams of humid regions seem to show much less variability. . .'. Thus the obtained results cannot be generalized to all climatic zones at present, because they pertain to a sample taken from rivers of the temperate climate only, in its transitional and arid continental versions. It should be remembered, however, that this is the zone that most palaeohydrological works deal with when discussing the problem of channel pattern changes in time and the hydrological significance of the channel pattern. For this reason, the results of the testing of the hypothesis about a relation between the channel pattern and the hydrologic regimen of a river justify some general thoughts.

1. It seems that the lack of any connection between the channel pattern and the spectrum of discharges will prove to be a general feature. The hydrologic regimen of a river cannot change several times along its course, especially if the river has no tributaries with a regimen decidedly different from its own. If we accept FALKOWSKI's hypothesis (1971, 1975, 190), we would need one single channel pattern along the river. Numerous examples show, however, that the channel pattern can change several times along the course of a river (LEOPOLD & WOLMAN 1957; LEOPOLD, WOLMAN & MILLER 1964; BRICE 1964; RUSSEL 1954; FALKOWSKI 1971, 1975). RUSSEL (1954) draws attention to the fact that the Meander River in Turkey from which the term 'meandering' comes has braided, meandering and straight reaches. The present authors know rivers in Mongolia, such as the Olgoin-gol and the Tsagan-turutuin-gol in the Khangai Mountains, which change their channel pattern a few times along their course.

2. Observations of recent channel processes and experiments in flumes show that the channel pattern of a stream '. . . is controlled by the mutual interaction of a number of variables, and the range of these variables in nature is continuous. . .' (LEOPOLD & WOLMAN 1957: 59). According to LANE (1957), these variables are: the discharge, gradient, sediment load, resistance of bed and bank to erosion, vegetation, geology, temperature, and human agency. BRICE (1964) is of the opinion that differences in the channel for and channel pattern can easily be explained by differences in the gradient, bed-bank material, and mean discharge, with bank erodibility being the most important single variable, depending mainly on the particle size of bank material and on the vegetation growing along the banks. FRIEDKIN (1945) emphasizes the importance of the speed of lateral erosion in the formation of the channel and its pattern. If the resistance of the banks is very small, the result is a braided channel. A low thre-

shold of bank erosion and the bed load transport are the fundamental conditions of braiding according to LEOPOLD, WOLMAN & MILLER (1964). Also vegetation influences the boundary value of this threshold. If its changes from thick to thin along the course of a river, or disappears completely, the channel changes from non-braided to braided. In the opinion of these scholars, '. . . the existence of one or another pattern is closely related to the amount and character of available sediment and to the quantity and variability of the discharge' (LEOPOLD et al. 1964: 281). However, as to the role of discharge variability, the authors state that 'Rapidly fluctuating changes in stage contribute to the instability of the transport regime and to erosion of the banks; hence they also provide a contributory but not essential element of the braiding environment' (LEOPOLD et al. 1964: 294).

From the above, we can infer the basic and direct causes producing specific types of channel pattern and its changes. The channel pattern is a result of responses between flow velocity and bed and bank erodability. The latter is a derivative of: particle size, and hence cohesion, the thickness of vegetation and its root system, and the heterogeneity of material. Erodability is the regulator of the flow velocity-material transport system.

3. The differentiation of the direct causes of channel pattern changes from indirect ones will allow us to see, in the right perspective, the role of the climatic factor on the one hand and the properties of the river basin as a local factor on the other, in the formation of the channel shape and pattern. This is shown on a model of the system forming the channel pattern (Fig. 10). Flow velocity depends on the gradient, which is a property of the basin, on the discharge volume, which is a derivative of the climate, and on roughness, which follows both from the properties of the drainage basin particle size and the climatic features the type and thickness of vegetation along the river banks. Also, both types of factors determine erodability as a regulator of the flow velocity-material transport system. A change in climatic conditions changes flow velocity through a change in the discharge and vegetation. Since flow velocity only partly depends on the climatic-hydrological factor, a definite change in climatic conditions need not produce the same results in two different channels if the remaining local factors affecting flow velocity and the erodability threshold are different in the two river basins. Hence, the responsiveness of rivers to climatic change understood as the formation of a specified channel pattern will be different, too. It follows that, in some cases, slight changes in the climatic conditions can be accompanied by bigger changes in the channel pattern than in other ones in which the same or more important changes in the climate do not result in channel pattern changes. In the first case, the erodability threshold has been crossed in one of the two possible directions, while in the other it has not been reached at all (Fig. 11).

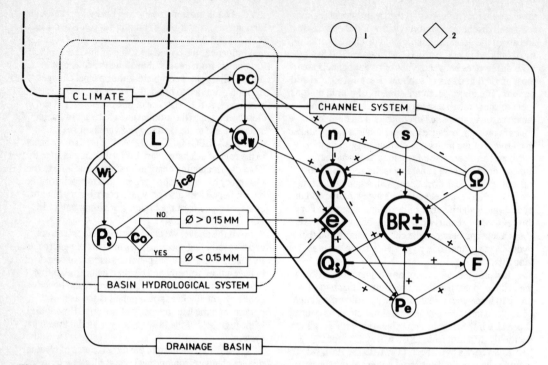

Figure 10. Model of the system forming the type of channel pattern – relationships between the factors controlling river channel pattern.

An arrow denotes the suggested direction of control, a plus denotes direct relationship, a minus is inverse relationship; the factors controlling the type of channel pattern directly are indicated with a thick line; braiding[+] denotes a change to the braided channel pattern, braiding[–] denotes a change to the meandering channel pattern. 1, variables; 2, regulators; Symbols of variables and regulators: BR, braiding; v, flow velocity; Q_s, sediment discharge; Pe, perimeter; F, width/depth ratio; Ω, sinuosity; s, gradient; n, roughness; Q_w, water discharge; Ps, particle size in the channel bed and banks; L, lithology; PC, density of plant cover; wi, weathering and its type; ica, infiltration capacity; e, erodability; co, cohesion.

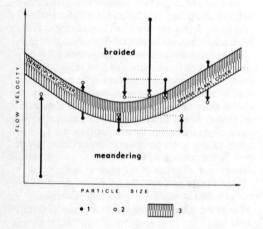

Figure 11. Changes of flow velocity under the influence of changes of climatic conditions or other physiographic properties of the drainage basin, and the erodability threshold and channel pattern.

As arrow denotes the flow-velocity change and its magnitude, black circle represents an initial flow velocity, white circle denotes a flow velocity after the change of climatic conditions or other physiographic properties of the drainage basin.

220

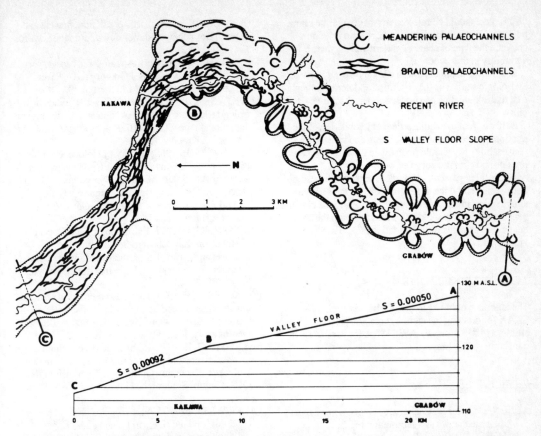

Figure 12. Change of the Holocene channel pattern on the valley floor along the course of the Prosna near Kakawa south of Kalisz B-B', section from which there is a change of the longitudinal gradient of the valley floor; A-B-C, longitudinal profile of the valley floor.

4. If the important local influence on the channel pattern formation is considered, how can the transformation from the braided to the meandering channel at the turn of the last cold period and the Holocene in the temperate climatic zone be explained? There are several aspects – the first is a similarity in the lithology and relief of many examined river basins of this climatic zone (the North European Plain). The second aspect is the tremendous climatic changes about 10,000 years ago which caused, even in basins with varying lithological properties and relief, the flow velocity to drop below the threshold of easy bank erodability, which led to river meandering. There is a possible third aspect: the conclusions on the change from braided to meandering channels at the turn of the last glacial and the Holocene are based on investigations of a relatively small number of valleys, and these tend to be valleys with particularly well preserved traces of old meandering channel pattern. It is not surprising that such examples are preferred test-sites (ROTNICKI &

LATAŁOWA 1986). This is why different channel pattern responses to the climatic changes may not have been discovered yet, as recent findings demonstrate. For instance TOMCZAK (1982) has established that meandering channels did not develop in the Vistula valley between Toruń and Solec Kujawski in the Holocene, although they did develop in other reaches of the river. Therefore, her conclusion that channels do not always respond to climatic changes in the same way. Recently, the present authors have found that some reaches of the Prosna did meander in Holocene times, while its lower course was braided from the point of the increased gradient of the valley floor (Fig. 12). It began, only a short time ago, to meander also.

9 GENERAL CONCLUSIONS

1. The hypothesis that the channel pattern is controlled by the hydrologic regimen, understood as a spectrum of

221

discharges, should be rejected for the temperate climatic zone. Research implies that there is no relation between the type of channel pattern and discharge variability in this zone.

2. Thus the channel pattern cannot be used in palaeohydrologic studies as the criterion for determining a past hydrologic regimen understood as discharge variability.

3. If we want to acquire a deeper knowledge of the relation between the channel pattern and river discharge variability, we should investigate the issue for other regions and climatic zones as well.

4. Useful in palaeohydrological terms are investigations which allow the determination, for a given type of alluvia and a given type of vegetation along the channel banks, of those threshold velocities at which the channel pattern changes from braided to meandering and vice versa.

ACKNOWLEDGEMENTS

The authors would like to thank Miss Gabriela PIŃ-CZAK for drawing the eight figures presented in this article, and Miss Maria KAWIŃSKA for her efforts in translating it into English.

REFERENCES

ALLEN, J.R.L. 1977. Changeable rivers: some aspects of their mechanics and sedimentation. In K.J.GREGORY (ed.), River Channel Changes, Wiley, Chichester, 15-45.

ANDRZEJEWSKI, L. 1984. The Zgłowiączka valley – its origin and development in the Late Goacial age and Holocene (Summary). Polish Academy of Sci., Institute of Geography and Spatial Organization, Fasc. 3, 84 pp.

BRICE, J.C. 1964. Channel patterns and terraces of the Loup Rivers in Nebraska. United States Geological Survey Professional Paper 422-D, 41 pp.

CHORLEY, R.J. & B.A.KENNEDY 1971. Physical Geography. A System Approach. Prentice-Hall International Inc., London, 370 pp.

DĘBSKI, K. 1954. Probability of hydrological and meteorological phenomena. The Method of deciles (in Polish). Wyd. Komunikacji, Warszawa, 83 pp.

DĘBSKI, K. 1970. Hydrology (in Polish). Ed. Arkady, Warszawa, 368 pp.

DURY, G.H. 1964a. Principles of underfit streams. United States Geological Survey Professional Paper 452-A, Washington.

DURY, G.H. 1964b. Subsurface exploration and chronology of underfit streams. United States Geological Survey Professional Paper 452-B, Washington.

DURY, G.H. 1965. Theoretical implications of underfit streams. United States Geological Survey Professional Paper 452-C, Washington.

DURY, G.H. 1976. Discharge prediction, present and former, from channel dimensions. Journal of Hydrology, 30, 219-245.

DURY, G.H. 1977. Underfit streams: retrospect, perspect, and prospect. In K.J.Gregory (ed.), River Channel Changes, Wiley, Chichester, 281-293.

DYNOWSKA, J. 1972. Types of River Regimes in Poland (Summary). Zeszyty Naukowe Uniwersytetu Jagiellońskiego CCLXVIII, Prace Geograficzne, Zeszyt 28, Kraków, 155 pp.

FALKOWSKI, E. 1965. History of Holocene and prediction of evolution of Middle Vistula valley from Zawichost to Solec. Materiały Sympozjum: Geologiczne problemy zagospodarowania Wisły środkowej od Sandomierza do Puław, SIT Górn. Zarząd Główny, Katowice, 45-61.

FALKOWSKI, E. 1967. Evolution of the Holocene Vistula from Zawichost to Solec with an engineering-geological prediction of further development (Summary). Biuletyn Instytutu Geologicznego, 198, 57-150.

FALKOWSKI, E. 1971. History and prognosis for the development of bed configurations of selected sections of Polish Lowland rivers (Summary). Biuletyn Geologiczny, 12, 5-110.

FALKOWSKI, E. 1972. Regularities in development of lowland rivers and changes in river bottoms in the Holocene. Excursion Guidebook, Symposium INQUA Comm. Stud. of the Holocene, 2, PAN, Poland, 3-30.

FALKOWSKI, E. 1975. Variability of channel processes of lowland rivers in Poland and changes of the valley floors during the Holocene. Biuletyn Geologiczny, 19, 45-78.

FALKOWSKI, E. 1980. Zasady ustalania schematycznego profilu geologicznego. . . Przegląd Geologiczny, 9, 496-501.

FALKOWSKI, E. 1982. The pattern of changes in the middle Vistula valley floor. In L.STARKEL (ed.), Evolution of the Vistula River Valley During the Last 15,000 Years, Part I, Geographical Studies, Special Issue, 1, PAN, 79-92.

FRIEDKIN, J.F. 1945. A laboratory study of the meandering of alluvial rivers. US Waterways Experiment Station, Vicksburg, Mississippi, 1-19.

FROEHLICH, W., L.KASZOWSKI & L.STARKEL 1977. Studies of present-day and past river activity in the Polish Carpathians. In K.J.Gregory (ed.), River Channel Changes, Wiley, Chichester, 411-428.

GREGORY, K.J. 1977. The context of river channel changes. In K.J.GREGORY (ed.), River Channel Changes, Wiley, Chichester, 1-12.

GREGORY, K.J. (ed.) 1977. River Channel Changes, Wiley, Chichester, 448 pp.

GREGORY, K.J. (ed.) 1983. Background to Palaeohydrology. A Perspective. Wiley, Chichester, 486 pp.

HEINE, K. 1982. Das Mündungsgebiet der Ahr im Spät-Würm und Holozän. Erdkunde, Band 36, Heft 1, 1-11.

HICKIN, E.J. 1977. The analysis of river-planform responses to changes in discharge. In K.J.Gregory (ed.), River Channel Changes, Wiley, Chichester, 249-263.

HITCHCOCK, D. 1977. Channel pattern changes in divided reaches: An example in the course bed material of the Forest of Bowland. In K.J.GREGORY (ed.), River Channel Changes, Wiley, Chichester, 207-220.

KLIMEK, K. & L.STARKEL 1974. History and actual tendency of flood-plain development at the border of the Polish Carpathians. In Geomorphologische Prozesse und Prozesskombinationen in der Gegenwart unter verschieden Klimabedingungen. Abhandlungen der Akademie der Wissenschaften in Göttingen, 185-196.

KNIGHTON, A.D. 1977. Short-term changes in hydraulic geometry. In K.J.GREGORY (ed.), River Channel Changes, Wiley, Chichester, 101-119.

KNOX, J.C. 1982. Responses of river systems to holocene climates. In Late-Quaternary Environments of the United states, vol. 2. The Holocene, H.E.WRIGHT Jr. (ed.), Minneapolis, University of Minnesota Press, 26-41.

KOZARSKI, S. 1974a. Late glacial and Holocene changes of river channel patterns in the lowland part of the Odra drainage basin in Polish. Krajowe Sympozjum 'Rozwój den dolinnych . . . etc.', Wrocław-Poznań, Streszczenie referatów i komunikatów, 17-19.

KOZARSKI, S. 1974b. Site Jaszkowo near Srem. Migrations of the Warta channel to the south of Poznań during Late glacial and Holocene – Generations of Meanders in Polish. Krajowe Sympozjum 'Rozwój den dolinnych . . . etc.', Wrocław-Poznań, Przewodnik wycieczki, 46-49.

KOZARSKI, S. 1981. River channel changes in the Warta valley to the south of Poznań. In S.KOZARSKI & K.TOBOLSKI (eds), Symposium 'Palaeohydrology of the temperate zone', Poznań, Poland '81, Guidebook of excursions, 6-23.

KOZARSKI, S. 1983a. River channel changes in the middle reach of the Warta Valley, Great Poland Lowland. Quaternary Studies in Poland, 4, 159-169.

KOZARSKI, S. 1983b. River channel adjustment to climatic change in west central Poland. In K.J.GREGORY (ed.), Background to Palaeohydrology, Wiley, Chichester, 355-374.

KOZARSKI, S. & K.ROTNICKI 1977. Valley floors and changes of river pattern channel in the North Polish Plain during the Late Würm and Holocene. Quaestiones Geographicae, 4, 51-93.

KOZARSKI, S. & K.ROTNICKI 1983. Changes of river channel pattern and the mechanism of valley-floor construction in the North-Polish Plain during the Late Weichsel and Holocene. In D.J.BRIGGS & R.S.WATERS (eds), Studies in Quaternary Geomorphology, Cambridge, 31-48.

KUZIN, P.S. 1955. Rivers (in Russian). BSE, 36, Moscow.

KUZIN, P.S. 1960. River Classification and Hydrologic Regionalisation of the USSR (in Russian). Gidromet Izdat., Leningrad, 454 pp.

LANE, E.W. 1957. A study of the shape of channels formed by natural streams flowing in erodable material. US Army Corps of Engineers, Missouri River Div., Omaha, Nebr., Sediment Ser. 9, 106 pp.

LEOPOLD, L.B. & M.G.WOLMAN 1957. River channel patterns – braided, meandering and straight. United States Geological Survey Professional Paper 282-B, 85 pp.

LEOPOLD, L.B., M.G.WOLMAN & J.P.MILLER 1964. Fluvial Processes in Geomorphology. W.H.Freeman & Co., San Francisco and London, 522 pp.

LEWIN, J. 1977. Channel pattern changes. In K.J.GREGORY (ed.), River Channel Changes, Wiley, Chichester, 167-184.

LEWIN, J. 1978. Meander development and floodplain sedimentation: a case study from mid-Wales. Geol. J., 13, 25-36.

LEWIN, J. 1983. Changes of channel patterns and floodplains. In K.J.GREGORY (ed.), Background to Palaeohydrology, Wiley, Chichester, 303-319.

MAIZELS, J.K. 1983a. Channel changes, palaeohydrology and deglaciation: Evidence from some Late Glacial sandur deposits of Northeast Scotland. Quaternary Studies in Poland, 4, 171-187.

MAIZELS, J.K. 1983b. Palaeovelocity and palaeodischarge determination for coarse gravel deposits. In K.J.GREGORY (ed.), Background to Palaeohydrology, Wiley, Chichester, 102-139.

MYCIELSKA-DOWGIAŁŁO, E. 1972. Stages of Holocene evolution of the Vistula valley on the background of its older history, in the light of investigations carried out near Tarnobrzeg. Excursion Guidebook, Symposium INQUA Comm. Stud. of the Holocene, 2, PAN, Poland, 69-82.

MYCIELSKA-DOWGIAŁŁO, E. 1977. Channel pattern changes during the last glaciation and Holocene, in the northern part of the Sandomierz Basin and the middle part of the Vistula valley, Poland. In K.J.GREGORY (ed.), River Channel Changes, Wiley, Chichester, 75-87.

PARDÉ, M. 1949. Fleuves et Riviéres. Sur les régimes de quelques gros fleuves. Lyon.

PARK, C.C. 1977. Man-induced changes in stream channel capacity. In K.J.GREGORY (ed.), River Channel Changes, Wiley, Chichester, 121-144.

ROSE, J., G.TURNER, G.R.COOPE & M.D.BRYAN 1980. Channel changes in a lowland river catchment over the last 13,000 years. In R.A.CULLINGFORD, D.A.DAVIDSON & L.LEWIN (eds), Timescales in Geomorphology, Wiley, Chichester, 159-175.

ROTNICKA, J. 1976. The separation of hydrological periods and description of river regimes by compar.-

son of probability distribution of water stages by pentads. Quaestiones Geographicae, 3, 79-102.

ROTNICKA, J. 1977. Theoretical Basis for Distinction of Hydrological Periods and for Analysis of River Regime Exemplified on the Prosna River. PTPN, Prace Komisji Geograficzno-Geologicznej, 18, Poznań, 94 pp.

ROTNICKI, K. 1974. Site Mirków. Stratigraphy of the Holocene deposits and main fluvial tendencies in the Prosna valley during the Holocene in Polish. Excursion Guidebook, Krajowe Sympozjum 'Rozwój den dolinnych...', Wrocław-Poznań, 49-55.

ROTNICKI, K. 1983a. Modelling past discharges of meandering rivers. In K.J.GREGORY (ed.), Background to Palaeohydrology, Wiley, Chichester, 321-354.

ROTNICKI, K. 1983b. Changes in the discharge of meandering rivers in North Polish Plain in the Late Glacial and Holocene. In K.E.BARBER & K.J.GREGORY (eds), Abstracts of Papers Severn 1983, IGCP Project 158 Palaeohydrology of the temperate zone in the last 15,000 years, Severn 1983, Symposium in UK, 19-26 September 1983, 76-78.

ROTNICKI, K. & M.LATAŁOWA 1986. Palaeohydrology and fossilization of a meandering channel of Younger Dryas age in the middle Prosna river valley. Quaternary Studies in Poland, 6 (in print).

RUSSEL, R.J. 1954. Alluvial morphology of Anatolian rivers. Ass. Amer. Geogr. Ann., 44, 363-391.

SCHUMM, S.A. 1960. The shape of alluvial channels in relation to sediment type. United States Geological Survey Professional Paper 352-B, Washington.

SCHUMM, S.A. 1963. Sinuosity of alluvial rivers on the Great Plains. Geological Society of America Bulletin, 74, 1089-1100.

SCHUMM, S.A. 1965. Quaternary palaeohydrology. In H.E.WRIGHT & D.G.FREY (eds), Quaternary of the United States, Princeton University Press, 783-794.

SCHUMM, S.A. 1968. River adjustment to altered hydrologic regimen – Murrumbidgee River and palaeochannels. Australia. United States Geological Survey Professional Paper 598, Washington.

SCHUMM, S.A. 1969. River metamorphosis. Journal of the Hydraulics Division, 95, Proceedings of the American Society of Civil Engineers, 251-273.

SCHUMM, S.A. 1971. Fluvial geomorphology: channel adjustments and river metamorphosis. In H.W.SHEN (ed.), River Mechanics, Vol. I, Fort Collins, Colorado, 5-22.

SCHUMM, S.A. 1974. Geomorphic thresholds and complex response of drainage systems. In M.MORISAWA (ed.), Fluvial Geomorphology, State Univ. of New York, Binghampton, New York, 299-310.

STARKEL, L. 1977. Palaeogeography of the Holocene (in Polish). PWN, Warszawa, 362 pp.

STARKEL, L. (ed.) 1981. The evolution of the Wisłoka Valley near Dębica during the last glacial and Holocene. Folia Quaternaria, 53, Ossolineum, Kraków, 91 pp.

STARKEL, L. 1983. The reflection of the hydrological changes in the fluvial environment of the temperate zone during the last 15,000 years. In K.J.GREGORY (ed.), Background to Palaeohydrology, Wiley, Chichester, 213-235.

STARKEL, L., K.KLIMEK, K.MAMAKOWA & E.NIEDZIAŁKOWSKA 1982. The Wisłoka river valley in the Carpathian Foreland during the Late Glacial and the Holocene. In L.STARKEL (ed.), Evolution of the Vistula River Valley During the Last 15,000 Years, Part I, Geographical Studies, Special Issue No. 1, PAN, Wrocław-Warszawa, 41-56.

SZUMAŃSKI, A. 1972. Changes in the development of the Lower San channel pattern in the Late Pleistocene and Holocene. Excursion guidebook, Symposium INQUA Comm. Stud. of the Holocene, 2, PAN, Poland, 55-69.

SZUMAŃSKI, A. 1982. The evolution of the Lower San river valley during the Late Glacial and the Holocene. In L.STARKEL (ed.), Evolution of the Vistula River Valley During the Last 15,000 Years, Part I, Geographical Studies, Special Issue No. 1, PAN, Wrocław-Warszawa, 57-78.

THORNES, J.B. 1977. Hydraulic geometry and channel change. In K.J.GREGORY (ed.), River Channel Changes, Wiley, Chichester, 91-100.

TOMCZAK, A. 1982. The evolution of the Vistula river valley between Toruń and Solec Kujawski during the Late Glacial and the Holocene. In L.STARKEL (ed.), Evolution of the Vistula River During the Last 15,000 Years, Part I, Geographical Studies, Part I, Special Issue No. 1, Wrocław-Warszawa, 109-129.

WIŚNIEWSKI, E. 1982. The geomorphological evolution of the Vistula river valley between Włocławek and Ciechocinek during the last 15,000 years. In L.STARKEL (ed.), Evolution of the Vistula River Valley During the Last 15,000 Years, Part I, Geographical Studies, Special Issue No. 1, PAN, Wrocław-Warszawa, 93-107.

Lake, Mire and River Environments, Lang & Schlüchter (eds)
© 1988 Balkema, Rotterdam. ISBN 90 6191 849 9

List of participants

Ammann, Brigitta, Dr.
   Systematisch-Geobotanisches Institut, Universität Bern, Altenbergrain 21,
   CH-3013 Bern, Switzerland

Ammann, Klaus, Dr.
   Systematisch-Geobotanisches Institut, Universität Bern, Altenbergrain 21,
   CH-3013 Bern, Switzerland

Astaras, Theodoros, Prof.Dr.
   Dept. of Geology and Physical Geography, Aristotle University Thessaloniki,
   GR-54006 Thessaloniki, Greece

Baltackov, Georgi, Dr.
   Dept. of Geomorphology and Cartography, Sofia University, 15 Russki BD,
   BG-1000 Sofia, Bulgaria

Beaulieu, Jacques-Louis, de, Dr.
   Lab. de Bot.Hist. et Palynologie, Faculté d. Sciences et Techn. de Saint-Jérôme,
   Rue Henri Poincaré, F-13397 Marseille, Cedex 1, France

Behre, Karl-Ernst, Prof. Dr.
   Niedersächs. Landesinstitut f. Marschen- und Wurtenforschung, Viktoriastr. 26/28,
   D-2940 Wilhelmshaven, FRG

Berglund, Björn E., Prof.Dr.
   Dept. of Quaternary Geology, Tornavägen 13, S-22 363 Lund, Sweden

Birks, John, Dr.
   Botanical Institute, University of Bergen, P.O.Box 12, N-5014 Bergen, Norway

Bortenschlager, Inez, Dr.
   Institut für Botanik, Sternwartestr. 15, A-6020 Innsbruck, Austria

Bortenschlager, Sigmar, Prof.Dr.
   Institut für Botanik, Sternwartestr. 15, A-6020 Innsbruck, Austria

Boucherle, M.M., Dr.
   Dept. of Biology, Queen's University, CDN-K7L 3N6 Kingston, Ontario, Canada

Bozilova, Elisaveta., Dr.
   Faculty of Biology, University of Sofia, Bulvar Russki 15, BG-1040 Sofia, Bulgaria

Brande, Arthur, Dr.
   Institut für Oekologie TU, Schmidt-Ott-Str. 1, D-1000 Berlin 41

Burga, Conradin, Dr.
   Geographisches Institut, Universität Zürich-Irchel, Winterthurerstr. 190,
   CH-8057 Zürich, Switzerland

Coccolini, Gemma, Dr.
   Via Svezia 9, I-00196 Roma, Italy

Chaix, Louis, Dr.
  Muséum d'Histoire Naturelle, 1 Route Malagnou, Case postale 284, CH-1211 Genève, Switzerland

Coûteaux, Michel, Dr.
  Laboratoire de Botanique historique et Palynologie, Rue Henri Poincaré, F-13397 Marseille Cedex 13, France

Delcourt, Hazel R., Dr.
  Dept. of Botany, University of Tennessee, Knoxville, Tennessee 37996, USA

Delcourt, Paul A., Dr.
  Dept. of Geological Sciences, University of Tennessee, Knoxville, Tennessee 37996, USA

Dickson, James, H., Dr.
  Botany Department, The University, Glasgow G12 8QQ, U.K.

Eicher, Ueli, Dr.
  Dennliweg 55, CH-4900 Langenthal , Switzerland

Elias, Scott, Dr.
  Instaar, Campus Box 450, University of Colorado, Boulder, Co 80309, USA

Fäh, Josef
  Systematisch-Geobotanisches Institut, Universität Bern, Altenbergrain 21 CH-3013 Bern, Switzerland

Gaillard, Marie-José, Dr.
  Kvartärbiologiska laboratoriet, Tornavägen 13, S-22363 Lund, Sweden

Ghenea, Constantin, Prof.Dr.
  Institute of Geology and Geophysics, Str. Caransebes 1, R-78344 Bucuresti, Rumania

Hadorn, Philippe
  Systematisch-Geobotanisches Institut, Universität Bern, Altenbergrain 21, CH-3013 Bern, Switzerland

Havlicek, Pavel, Dr.
  Geological Survey, Malostranské nam. 19, CS-118 21 Praha 1, Czecho-Slovakia

Hicks, Sheila, Dr.
  Dept. of Geology, University of Oulu, Linnanmaa, SF-90570 Oulu 57, Finland

Hofmann, Wolfgang, Dr.
  Max-Planck-Institut für Limnologie, Postfach 165, D-2320 Plön, FRG

Hubschmid, Felix
  Pestalozzistrasse 1, CH-3400 Burgdorf, Switzerland

Huybrechts, Willy, Dr.
  Geografisch Instituut, Vrije Universiteit Brussel, Pleinlaan B-1050 Brussel, Belgium

Huysmans, Linda
  Paardenkerkhofstr. 154, B-2800 Mechelen, Belgium

Jacomet, Stefanie, Dr.
  Botanisches Institut der Universität Basel, Schönbeinstrasse 6, CH-4056 Basel, Switzerland

Johansen, Johannes, Dr.
  Museum of Natural History, DK-3800 Torshavn, Faroe Islands

Kordos, Làszlò, Dr.
  Hungarian Geological Institute, Népstadion u. 14, H-1143 Budapest, Hungary

Koutaniemi, Leo, Dr.
  Dept. of Geography, University of Oulu, SF-90570 Oulu, Finland

Kozarski, Stefan, Prof.Dr.
  Quaternary Research Institute UAM, ul. Fredry 10, PL-61-701 Poznan, Poland

Lang, Gerhard, Prof.Dr.
  Systematisch-Geobotanisches Institut der Universität Bern, Altenbergrain 21,
  CH-3013 Bern, Switzerland

Latałowa, Malgorzata, Dr.
  Dept. of Plant Ecology and Nature Protection, University of Gdansk, Czolgistow 46,
  PL-81-378 Gdynia, Poland

Lemdahl, Geoffrey
  Kvartärbiologiska Laboratoriet, Tornavägen 13, S-22363 Lund, Sweden

Lotter, André
  Systematisch-Geobotanisches Institut, Universität Bern, Altenbergrain 21,
  CH-3011 Bern, Switzerland

Magny, Michel, Dr.
  Laboratoire de Chrono-Ecologie, Faculté des Sciences, La Bouloie,
  F-25030 Besançon Cedex, France

Marciniak, Barbara, Dr.
  Inst. Nauk Geologicznych PAN, Al. Zwirki i Wigury 93, PL-02-089 Warszawa, Poland

Merkt, Josef, Dr.
  Nds. Landesamt für Bodenforschung, Postfach 510153, D-3000 Hannover-51, FRG

Mike, Karoly, Dr.
  Research Centre for Water Resource Development, Vituki, Kvassay Jeno ut 1,
  H-1095 Budapest, Hungary

Miller, Urve, Dr.
  Institute of Quaternary Research, University of Stockholm, Odengatan 63,
  S-113 22 Stockholm, Sweden

Moe, Dagfinn, Dr.
  Botanical Institute, University of Bergen, P.O.Box 12, N-5014 Bergen, Norway

Moulin, Bernard
  Fouilles de Champreveyres, rue des Roseaex, CH-2068 Hauterive, Switzerland

Mørkved Vorren, Brynhild
  Tromsö Museum, University of Tromsö, N-9000 Tromsö, Norway

Müller, Helmut, Dr.
  Bundesanstalt für Geowissenschaften und Rohstoffe, Postfach 510153,
  D-3000 Hannover-51, FRG

Müller, Hans-Niklaus, Dr.
  Geographisches Institut, Justus-Liebig Universität, Senkenbergstr. 1,
  D-6300 Giessen, FRG

Oeggl, Klaus, Mag.
  Botanisches Institut, Universität Innsbruck, Sternwartestrasse 15,
  A-6020 Innsbruck, Austria

Okuniewska, Iwona, Mag.
  Wielkopolska Archaeology Dept., Inst. of Material Culture History,
  ul. Zwierzyniecka 20, PL-60-814 Poznan, Poland

Paus, Aage
  Botanical Inst. Universitet, Box 12, N-5014 Bergen, Norway

Radai, Ödön, Dr.
  Vituki, Kvassay Jeno ut 1, H-1095 Budapest, Hungary

Ralska-Jasiewiczowa, Magdalena, Dr.
  Zaklad Paleobotanyki Inst. Bot. PAN, ul. Lubicz 46, PL-31-512 Krakow, Poland

Raukas, Anto, Prof.Dr.
  ENSV TA Geologia Instituut, Estonia pst. 7, USSR-200101 Tallinn-1, USSR

227

Ritchie, James C., Prof.Dr.
   Scarborough College, University of Toronto, 1265 Military Trail,
   Scarborough, ONT. MIC 1A4, Canada

Rotnicki, Karol, Prof.Dr.
   Quaternary Research Institute UAM, Fredry 10, PL-61-701 Poznan, Poland

Rybnickova, Eliska, Dr.
   Inst. of Experiment. Phytotechnology, Czechoslovak Academy of Sciences, Stara 18,
   CSSR-662 61 Brno, Czecho-Slovakia

Saarse, Leili, Dr.
   ENSV TA Geoloogia Instituut,Estonia pst. 7, USSR-200101 Tallinn-1, USSR

Schirmer, Wolfgang, Dr.
   Abt. Geologie der Universität, Universitätsstrasse 1, D-4000 Düsseldorf, FRG

Schlüchter, Christian, Dr.
   Institut für Grundbau, ETH Hönggerberg, CH-8093 Zürich, Switzerland

Schneider, Ruth, Dr.
   Systematisch-Geobotanisches Institut, Universität Bern, Altenbergrain 21,
   CH-3013 Bern, Switzerland

Siegenthaler, Ulrich, Dr.
   Physikalisches Institut, Universität Bern, Sidlerstrasse 5, CH-3012 Bern,
   Switzerland

Smol, John, Dr.
   Dept. of Biology, Queen's University, Kingston Ontario, K7L 3N6, Canada

Starkel, Leszek, Prof.Dr.
   Inst. of Geography, Polish Acad. of Sc., ul. Sw. Jana 22, PL-31-018 Krakow, Poland

Straub, François
   Gymnase cantonale, CH-2300 La-Chaux-de-Fonds, Switzerland

Tobolski, Kazimierz, Prof.Dr.
   Quaternary Research Institute UAM, ul. Fredry 10, PL-61-701 Poznan, Poland

Tolonen, Kimmo, Prof.Dr.
   Dept. of Botany, University of Oulu, Linnanmaa, SF-90570 Oulu, Finland

Tolonen, Mirjami, Dr.
   Dept. of Botany, Ecological Laboratory, Fabianink. 24A, SF-001000 Helsinki 10,
   Finland

Tonkov, Spassimir
   Biological Faculty, University of Sofia, Bul. Russki 15, BG-1000 Sofia, Bulgaria

Tropeano, Domenico, Dr.
   CNR-Ist. Protezione Idrogeologica, Bacino Padano, Via Vassalli Eandi, 18,
   I-10138 Torino, Italy

Valsangiacomo, Antonio
   Systematisch-Geobotanisches Institut, Universität Bern, Altenbergrain 21,
   CH-3013 Bern, Switzerland

Vandenberghe, Jef, Dr.
   Inst. v. Aardwetenschappen, Vrye Universiteit, De Boelelaan 1085,
   NL-1081 Amsterdam HV, Netherlands

Vasari, Annikki, Mag.
   Dept. of Botany, University of Helsinki, Unioninkatu 44, SF-00170 Helsinki 17,
   Finland

Vasari Yrjö, Prof.Dr.
   Dept. of Botany, University of Helsinki, Unioninkatu 44, SF-00170 Helsinki 17,
   Finland

Verbruggen, Cyriel, Dr.
   Laboratorium voor Regionale Geografie en Landschapskunde, Rijksuniversiteit Gent,
   S8-Al Krijgslalan 281, 9000 Gent, Belgium

Vorren, Karl-Dag, Prof.Dr.
   University of Tromsö I.B.G., Postbox 3085, Guleng, N-9001 Tromsö, Norway

Wahlmüller, Notburga, Dr.
   Botanisches Institut, Universität Innsbruck, Sternwartestrasse 15,
   A-6020 Innsbruck, Austria

Waldis, Rolf,
   Systematisch-Geobotanisches Institut, Universität Bern, Altenbergrain 21,
   CH-3013 Bern, Switzerland

Watts, William A., Dr.
   Provost's House, Trinity College, EI - 2 Dublin, Ireland

Wegmüller, Samuel, Prof.Dr.
   Systematisch-Geobotanisches Institut, Universität Bern, Altenbergrain 21,
   CH-3013 Bern, Switzerland

Wick, Lucia
   Systematisch-Geobotanisches Institut, Universität Bern, Altenbergrain 21,
   CH-3013 Bern, Switzerland

Wohlfarth-Meyer, Barbara
   Labor für Urgeschichte, Petersgraben 9-11, CH-4051 Basel, Switzerland

Zoller, Heinrich, Prof.Dr.
   Botanisches Institut, Universität Basel, Schönbeinstrasse 6, CH-4056 Basel,
   Switzerland

Züllig, Hans, Dr.
   Brendenweg 9, CH-9424 Rheineck, Switzerland